DATE DUE		
APR 17 2000	MAY 2 2 2010	
JUL 25 2000	MAR 2 6 2012	
DEC 14 2000		
OCT 04 2003		
DEC 04 2004		
APR 1 2009		

Dirt

DIRT

JOHN
ANTHONY
ADAMS

Illustrations by
Jean Adams

TEXAS A&M
UNIVERSITY PRESS
COLLEGE
STATION

Manufactured in the United States of America
First Edition

Library of Congress Cataloging-in-Publication Data

Adams, John Anthony, 1944–
    Dirt.

    Includes index.
    1. Soils. I. Title.
S591.A63   1986   631.4   86-40213
ISBN 0-89096-272-3
ISBN 0-89096-301-0 (pbk.)

To the memory of

DR. NATHANIEL T. COLEMAN,

of the University of California at Riverside,

who interested me in becoming

a soil scientist

Contents

Illustrations

Preface

To the uninitiated soil may seem to be simple, static, and completely dull, posing the question "How could a book about soil be interesting?" The contrast between this view and the innumerable engrossing facts and ideas actually connected with this subject make dirt an appropriate topic for a book aimed at the adventurous reader, who stands to gain a fresh and often surprising view of a universal material. Let us take a brief preview of the chapters to show the range of topics to be explored.

Chapter 1 describes the enormous ignorance of people in past centuries about soil and considers why, in spite of the present great gains in knowledge, there remain so many misconceptions.

Chapter 2 looks at the wondrous ability of nature to turn rocks into sand, silt, and clay, and the reasons that each of these three types of soil particles behaves as it does. The discussion shows how the origin of land plants greatly changed the nature of the earth's primitive soils and examines some effects of animals on soils. The chapter compares the origin of soils in some of the most famous farmlands in the United States and points out how their fertility affected and was affected by the pioneering settlers.

In chapter 3 we discover what conditions prevent

the decay of organisms; for example, why are some ancient human bodies so well preserved? Disaster would befall all life if dead things did not decompose. After investigating what causes decay, this chapter surveys what goes on in a compost pile and concludes with a look at some composting practices that may do more harm than good.

Chapter 4 considers the nutrients that plants must get from the soil and contrasts the use of natural organic and synthetic fertilizers to supply these needs, thus leading to a discussion of the modern organic movement. The chapter then covers the origins of the vast amounts of nitrogen, phosphorus, and potassium used to make synthetic fertilizers, concluding with a demonstration of the impracticality of relying solely on natural organic fertilizers to maintain high yields.

The subject of chapter 5 is one that has many traps for unwary gardeners—maintaining the optimum amount of soil water. Why are some soils so maddeningly hard to wet, and why is it so easy to apply too much or too little water, even to absorbent soils? Chapter 5 describes a relatively simple and inexpensive instrument that can be used to maintain near-optimum soil moisture, and then takes a look at some misconceptions about watering, drainage, and drying. The discussion of soil water concludes by revealing the secrets of quicksand.

Chapter 6 explores the diverse ways that soil is worn away. Landslides, earthquake displacements, beach erosion, and dust bowls are all discussed. The good side of erosion—the creation of new lands by deposited sediments—also receives attention. The threat of erosion to U.S. farmlands is assessed, using the best data and projections from the U.S. Department of Agriculture.

Chapter 7 follows the illustrious history of the plow, showing how improvements in the design benefited humanity. After a review of the ignominy suffered by the plow in recent decades—the charge that it is responsible for huge losses of rich farmlands—the chapter evaluates the use of minimum tillage or non-tillage agriculture to reduce or eliminate the plow's harmful effects.

Soil classification is the subject of chapter 8, demonstrating how soils are named by combining meaningful syllables into nonsensical-sounding words. The arduous work of soil mappers is described and evaluated, and the uses and misuses of soil surveys are set forth—including the federal government's erroneous basis for grazing on public lands.

In chapter 9 we take a look at desert soils, particularly desert pavement and sand dunes, and at the amazing career of British soldier and sand dune expert Ralph Bagnold. The effects of recreational off-road driving on California desert soils are analyzed, and the possible effects of valley fever fungus on dust-raising desert recreationists is considered.

Chapter 10 explores one of the hottest current environmental controversies—acid rain, acid soils, and acid runoff. An examination of the origin of soil acidity shows that some of the most acid soils in the world occur *naturally* in areas where there is concern about acid rain effects. The buffering (resistance) of the soils to increases in acidity and the role of soil acidity in determining the acid content of runoff water are related. The discussion then considers the effect of reforestation in North America and some parts of Europe on soil and lake acidity. A survey of possible causes of forest damage follows.

Unusual soils are the focus of chapter 11. These include soils on the moon, on polders (lands reclaimed from the oceans), and in peat bogs, as well as tundra soils, tropical soils, and sands.

Preface

I am indebted to many people for their invaluable assistance in writing this book. Professor James L. Adams, my brother, a professor of mechanical engineering as well as chairman of the Values, Technology, Science, and Society Program at Stanford University, suggested writing a short book about soil entitled *Dirt*. He made many good editorial comments when the book was in its preliminary stages. Marian Player Adams, my sister-in-law, director of Continuing Education at Stanford, also made many helpful editorial suggestions during the early organization of the book.

Doctors Frank T. Bingham and Lewis H. Stolzy, professors of soil science at the University of California at Riverside (UCR), provided much invaluable help and reviewed the manuscript for accuracy. Other members of the Department of Soil Science at UCR provided helpful information in discussions, including Walt Farmer, Lanny Lund, Al Page, and Gary Sposito. I would like to thank Jean Adams, my mother, for her illustrations for the book.

Dirt

1
From Universal Ignorance to Widespread Misconceptions

Farmers have been growing crops for probably 9,000 years or longer, but over nearly all of that vast stretch of time they gained only a very tenuous understanding of soil. Even by the time of the American Revolution in the late eighteenth century, there was little accurate knowledge of plant needs, although there were plenty of imaginative theories. For example, Jethro Tull, an English agriculturist who invented a mechanical drill for sowing seed, had concluded that minute particles of soil were the proper "pablum" of plants. He believed that pressure caused by swelling roots forced these particles into the "lacteal mouths" of the roots, where they entered the plant's circulatory system. Tull thought that all plants lived on these particles.[1]

The most widely accepted ideas of the early nineteenth century—many of them totally wrong—were summarized by Sir Humphry Davy, a famous English chemist, in his *Elements of Agricultural Chemistry*, published in 1813. In general, Davy thought that plants ob-

---

1. Jethro Tull, *Horse Hoeing Husbandry: or an Essay on the Principles of Vegetation and Tillage. Designed to Introduce a New Method of Culture; Whereby the Produce of Land Will Be Increased and the Usual Expense Lessened Together with Accurate Descriptions and Cuts of the Instruments Employed in It*, 4th ed. (London: A. Millar, 1762), pp. 22, 8.

tained their carbon—the central element in the compounds of which organisms are made—through their roots. Thus he thought that oils were good manures because of their carbon and hydrogen content, and he concluded that soot was a fine fertilizer for plants because the carbon "is in a state in which it is capable of being rendered soluble by the action of oxygen and water."[2] Now, we know that essentially all carbon comes to plants from the carbon dioxide of the air, which green plants make into carbohydrates through the process of photosynthesis.

Davy's book shows other common misconceptions of the time. He thought that the value of applying lime to soil was that it would dissolve hard vegetable matter. Later it became clear that lime applications to soil are actually useful because they neutralize soil acidity. Davy also thought that once organic matter was dissolved, there was no advantage in letting it decompose further; therefore, he concluded wrongly that fresh urine was more useful as manure than putrid urine, and that it was a bad practice to allow farmyard manure to decompose before applying it to the land. Only later was it discovered that additional decomposition of organic matter will release more nutrients.

The American colonists, lacking good theories about soil fertility, did not develop improved farming techniques but instead adopted practices that had been used in Europe, many of them since Roman times. It was fortunate that most of the colonists had migrated from western Europe, where the soils were similar to those found in their new land.

Crop yields in the colonies were very low, as they had been in Europe. Colonial agriculture did manage

2. Humphry Davy, *Elements of Agricultural Chemistry in a Course of Lectures for the Board of Agriculture* (London: Longman, Hurst, Rees, Orme, and Brown, 1813), p. 268.

to produce wheat, tobacco, and rice in excess of domestic needs, allowing the colonists to trade the surplus to the French for guns to use against the British during the Revolution. But their agricultural practices were so poor that the soils became exhausted, and many farms were abandoned.

Primitive agricultural practices not only were hard on soil but were one of the factors that prevented more rapid economic growth in general. Backward farming techniques, including poor methods of soil management, kept output per capita (and therefore income per capita) at low levels. Since most people were farmers, the limited ability of farmers to make cash purchases hampered the early development of manufacturing.

Advances in chemistry, physics, and plant physiology starting in the late 1700s led to the emergence of soil science as a separate field of study in the 1890s and early 1900s. Since the 1940s, knowledge of the soil has grown rapidly, and the many new principles discovered have greatly improved practices of soil management.[3]

Included in the vast store of knowledge accumulated in the past several decades is information that is useful to gardeners, that can help to clarify some recent environmental controversies, or that is just interesting to know: the secret of quicksand, why some soils are almost impossible to wet, why clays behave as they do, the processes that can create 50-foot soil "blisters" on the tundra, methods that allow crop production on sand dunes, and on and on.

Much of this information is available only in technical or semitechnical publications that are either incom-

3. Two good sources for the history of soil science are E. W. Russell, *Soil Conditions and Plant Growth*, 10th ed. (London: Longman, 1973), ch. 1; and a series of articles by various authors in the *Soil Science Society of America Journal* 41, no. 2 (March-April, 1977): 221–65.

prehensible or unpalatable to the general reader. Even the majority of nonagricultural scientists have little understanding of soils, because as students they were either not inclined to enroll in agriculture-oriented courses or were trained at a college or university that did not offer such courses (e.g., Harvard, MIT, Stanford, Caltech, or UCLA).

The lack of training and of readable but accurate literature has resulted in perhaps more widespread misconceptions about soil than about any other important part of nature; they are evident in the erroneous concepts of soil held by many gardeners and also in the environmental controversies associated with soil. There are popularized books about wildlife, the atmosphere and weather, nuclear physics, geology, the oceans, astronomy, botany, and the like, but there have been no comparable books about soil, even though it is one of the most complex, fascinating, and indispensable parts of nature! This book is an attempt to fill that need. It discusses not only agricultural and gardening topics but many related nonagricultural matters: lunar soils, beach deposition and erosion, landslides, and even the relationship of soil to the preservation of human bodies.

The Storm

2
The Soil Factory

Primeval Soils

Wouldn't it be exciting if there were a window to the past that you could look through to see primordial landscapes? There is such a window for one part of those ancient scenes—soil surfaces. Although the ancient soils themselves no longer exist, their appearance is sometimes preserved where muds were turned into rock.

Shales (rocks formed mainly from consolidated clay) may show small pits made by raindrops that fell eons ago. Of even greater interest are the fortunate cases where animal tracks are beautifully preserved in rocks. Although most people think that bones are the main fossil evidence for extinct land animals, bones are actually rarer than tracks; most known extinct species have been discovered only by their tracks.

Tracks show evidence of the first known invasion of dry land by an animal with a backbone. These markings, found in 370-million-year-old rocks on the Orkney Islands, were made by a primitive fish when, propelled by stubby fins, it crawled on its belly from one pool of water to another.

Soils on which the dinosaurs trod give some fascinating clues to the behavior of those mysterious reptiles. For example, their footprints can give evidence of how fast the animal could move. By means of a formula based on the observed spacing and pattern of the prints, it has been estimated that one medium-sized carnivorous dinosaur traveled at a speed of ten miles per hour. Parallel sets of dinosaur tracks have suggested that carnivorous dinosaurs, at least those of small to medium size, hunted in packs. This evidence does not corroborate the long-standing view of dinosaurs as essentially solitary and slow moving.[1]

How long were there soils on earth before the first animals walked on them? The oldest rocks formed from sediments are about 3.8 billion years old. So we know that there have been soils for at least that length of time and probably much longer. Thus soils have existed on earth more than ten times as long as land animals with backbones. The soils that now cover the earth are not the primeval soils; they were long ago eroded or buried —often being compacted and cemented into sedimentary rocks.

The Origin of Soils

All soils originated from rocks. After the scorching hot crust of the earth began to cool billions of years ago, water—along with the action of its dissolved gases— began to weather and decompose the surface rocks. The surface was probably cool enough to allow water to lie on it about 4.45 billion years ago, so the weathering of rocks and formation of soil could have begun in earnest by then. Most geologists believe that the total area of the continents has increased with time. Estimates of

1. David J. Mossman and William A. S. Sarjeant, "The Footprints of Extinct Animals," *Scientific American* 248, no. 1 (January, 1983): 64–74.

the proportion of present continental material formed in different stages of the earth's past are as follows:

3.8 billion to 3 billion years ago—15 percent;
3 billion to 2.5 billion years ago—70 percent;
2.5 billion years ago to the present—15 percent.

It is interesting to note that a peak of continent building began about 2.8 billion years ago.

Therefore, the areas with a cover of soil increased dramatically during the earth's history. As continents and island arcs were pushed like flotsam around the earth's surface by the process of continental drift, these soil-covered areas occupied various locations on the earth's surface.

As rocks weathered, the more easily disintegrated rock minerals (such as feldspars and micas) formed clays, while the more resistant minerals (such as quartz) became sand or silt particles.

Many clays are made up of an enormous number of minute flat plates, or flakes, which cause their familiar properties.

1. Clays are slippery when wet because the plates slide over one another as would panes of glass with water film between them.

2. Clays become hard when dry because water molecules are lost between drying plates, allowing the particles to pack closely where strong attractive forces link them into a hard mass.

3. Clays swell when infiltrating water forces clay plates apart, and crack when the dry layers shrink back together—leaving fissures.

4. Clays hold water and nutrients because weak electrical charges in the clay particles (caused by imperfections in the structural arrangements of

KAOLINITE CLAY PARTICLES

atoms in clay crystals) attract nutrients and ends of water molecules with opposite electrical charges.[2]

Very wet clays can flow like a thick, viscous liquid. When saturated, they may begin to slip even on slight slopes. In eastern Canada and northwestern Europe, whole valley floors are layered with clay and sand that extend for miles inland and rise to elevations of several hundred feet. During spring thaws a mass of mud as large as a square mile may suddenly let go and flow downhill at speeds up to six miles an hour.

Tragedies have occurred when the silently sliding masses have engulfed sleeping communities during the night. Scores of farms in Trondheim, Norway, were destroyed by a mud flow on a May night in 1893; 111 people were drowned in mud, and other terrified residents escaped after riding down the valley in their farmhouses as far as three miles. In Saint Jean Vianney, Que-

2. A good discussion of clays is found in Nyle C. Brady, *The Nature and Properties of Soils*, 8th ed. (New York: Macmillan, 1974), pp. 71–110.

bec, on a May night in 1971, a spring thaw after record snows saturated clay under a housing development on a 100-foot cliff. Ninety million cubic yards of soil slid almost two miles before depositing houses, cars, sidewalks, and telephone poles in the Saguenay River; 31 people were killed and 40 houses were demolished.

Could your distant ancestor have been a rock? There has been an imaginative and interesting conjecture that clay has not only been of great importance to life but that life may have originated from clay particles. A Glasgow chemist, A. G. Cairns-Smith, has speculated that our first ancestors were literally made of clay. The DNA in cells provides instructions for their replication, but in the first living things perhaps no such complicated mechanism existed. Cairns-Smith suggests that the first means of replication may have been a defect pattern, spread across a clay crystal surface layer, which caused its own crude and approximate reproduction. These defects may have given a selective growth advantage to clay-based organisms, which developed competence in organic chemistry to gain greater control over their environment. Finally, Cairns-Smith submits, the new genetic material, DNA, evolved and took over the replicating function, and the clay particles were abandoned as life became established in cells.[3]

Cairns-Smith's arguments are developed ingeniously but remain entirely in the realm of speculation. He suggests that the possibilities could be tested by looking for the beginnings of the process in minerals today, by seeking organo-clay organisms as fossils, or by eventually finding out how to make some primary kinds of organisms in the laboratory. If it turns out that we are all

3. See A. G. Cairns-Smith, *Genetic Takeover and the Mineral Origins of Life* (Cambridge: Cambridge University Press, 1982).

descended from clay particles, some people will be sure to try to trace their ancestry back to rocks.

Some American Indians think of clay as alive. Joseph Lonewolf of the Santa Clara (New Mexico) Pueblo Indians is a famous potter, noted for his intricately incised and carved earthenware. Lonewolf said,

> You see, I believe the clay is a living thing because it comes from Mother Earth, a living thing. Without her, we'd not have food nor plants nor animals.
>
> I talk to her before I take the clay from her. And I believe the clay is a living thing and has feelings. I hurt the clay's feelings when I talk of a pot that hasn't been done yet. And the clay talks back by cracking and breaking. So I have learned to respect its feelings and never talk of a pot until it is completed—until it is finished and ready to talk for itself.[4]

Now that we have explored clay, it is time to consider the other important components of mineral soil—sand and silt. The discussion about these types of grains will be shorter, however, because they do fewer things in the soil; their main service is to counter the vices of clay in soil by providing larger pores between grains and therefore keeping the soil more open.

Silt and sand particles differ from clay particles in size. Whereas the great magnification of an electron microscope is needed to see individual clay particles, the larger silt particles can be seen with an ordinary light microscope; and sand particles, the largest of the three categories, are visible to the naked eye as irregular or rounded grains.

Both silt and sand are predominantly made of

4. Maggie Wilson, "The Beauty Makers," *Arizona Highways* 50, no. 5, (May, 1974): 37.

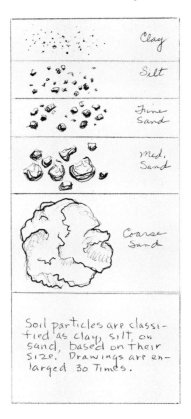

Soil particles are classified as clay, silt, or sand, based on their size. Drawings are enlarged 30 Times.

quartz. Properties of sand include that it (1) is very inert, (2) is not sticky even when wet and cannot be molded, (3) has low water- and nutrient-holding capacity, and (4) promotes good aeration and drainage.

Silt particles are essentially micrograins. Silt has these properties:

1. Silt, like sand, is more inert than clay but has some capacity to hold water and nutrients because the particles are often covered with clay.

2. It is more sticky and moldable than sand but much less so than clay.

3. Soils very high in silt create a thick choking dust.

Plants Enter

While rock minerals are weathering and being formed into clays, soluble elements such as potassium, calcium, and magnesium are released. These elements are essential for plant growth. The primeval soils of 3.8 billion years ago, however, were barren of any plant life, and soils remained bare until the land plants became established billions of years later.

When did land plants first take a foothold on the earth's soils and what kind of plants were they? An early clue was found by a pioneering paleobotanist named Sir John William Dawson. He was splitting rocks and examining the fragments on the Gaspé Peninsula of Canada in 1859 when he found a fossil land plant fragment that we now know was more than 350 million years old. Dawson's discovery was colorfully described by Donald Culross Peattie in his book, *Flowering Earth*.

> [Dawson] was a God-fearing, Bible-swearing gentleman who did not, in that year of grace 1859, take any stock in Mr. Darwin's blasphemy about the descent of man. But he was a good paleobotanist for all of that, and when he found a land plant square in the middle of the Age of Seaweeds, he knew he had made a discovery.
>
> He took his stony fragment home to Nova Scotia, where he was born, and went to work on it. Neither mad nor a magician, he dared to look back three hundred and fifty million years, and see what must have been growing then. He was so sure of what he saw that he could take up a pencil

and draw it. I have that picture before me. It is a picture of the earliest known plant upon the earth. Sir John called it Psilophyton, which means "naked plant." Very naked it looks, very new for all it is so old—a skinny, wiry, straggling thing, no more than the dim beginning of an idea for a plant. Which is just about what it was.[5]

The plant was about a foot tall, with some underground stem but no true roots. The stems bore spore cases at their tips (it was to be ages before seed-producing plants appeared). Peattie further observed, "This thing, this meagre, venturesome, growing, and certainly green thing, lost in the interminable darknesses of time gone by, came alive again in the mind of Sir John William Dawson."[6]

Dawson's colleagues thought that his imagination had run wild. For 50 years his claim was not taken seriously. Then, one day in 1915, two British paleobotanists were looking for fossils in mountains of Aberdeenshire when they unearthed an ancient marsh turned into rock and full of fossils. These fossils were close to the village of Rhynie, so Robert Kidston and William Lang gave the name Rhynia to the first of the plants they discovered there. They saw that Rhynia must have grown thickly in the bog like rushes in a marsh today. It had no leaves or roots but only underground stems and rootlets, and it bore spore cases.

Rhynia has since been found as a fossil at locations as far apart as Connecticut, Maine, Scotland, Wales, Germany, and Victoria in Australia. It is so close to Sir John William Dawson's drawing of Psilophyton that Psilophyton must have been a very real plant. As Peattie

5. Donald Culross Peattie, *Flowering Earth* (New York: Viking Press, 1961), pp. 105–106.
6. Ibid., p. 106.

17

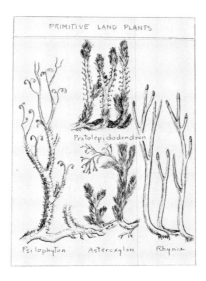

PRIMITIVE LAND PLANTS

Psilophyton   Asteroxylon   Rhynia

Protolepidodendron

stated: "No doubt any more of Dawson's bold guess, no doubt of the importance of Psilophyton, the 'naked plant,' the first known citizen in the land. Spores like a fern's give hint, in this bleak tentative little ancestor, of great things to come."[7]

Fossils of various extinct species of very early, simple land plants have since been found in rocks, and the earliest known appearance of vascular plants (those with conducting tissues for water and sap) has been traced to about 425 million years ago.[8]

Once vegetation developed, as though intelligently looking to its future, it greatly improved its own soil as a growing medium. Not only did rotted plant litter and its released nutrients make up a vital constituent of fertile soils, but the acids produced during decomposition of organic matter enhanced the breakdown of rock

7. Ibid., p. 107.

8. The evolution of early land plants is discussed in Wilson N. Stewart, *Paleobotany and the Evolution of Plants* (Cambridge: Cambridge University Press, 1983), pp. 53–83.

minerals. The intensified disintegration of minerals provided increased rates of clay formation and the release of soluble essential nutrients from minerals. The cover of vegetation further promoted the development and accumulation of soils by giving protection from erosion.

Soil Horizons

Plants have had a great deal to do with creating the layers in soil called soil horizons. Rotting litter darkens the top of the soil mass where it is mixed with mineral matter. Decaying organic matter in this upper layer releases plant nutrients and enhances breakdown of soil minerals, thereby accelerating the rate of clay formation and the release of more nutrients. These products may simply enrich the upper layers of soil or, where water consistently percolates down through the ground, may move down and accumulate as layers at a lower depth.

It is commonly assumed that dark-colored soils, with their greater amounts of organic matter, are more productive than light-colored soils. While this is generally true, some dark organic soils are so acidic that they are unproductive without heavy additions of lime. Other soils are dark in color merely because of the dark parent rock and may be poor for agriculture.

The natural tendency of soils to form slowly into differentiated layers is partly counteracted by the teeming underground work force of burrowing animals. Ants and termites provide two good examples of small creatures that can move a lot of ground.

Ants excavate colossal, intricately organized underground cities, which can extend horizontally for hundreds of feet and house a half million or more inhabitants. The earth removed from these diggings is piled

Soil layers (horizons) under New England conifer forest.

up in large mounds. Colonies of the eastern mound-building ant, *Formica exsectoides*, amass mounds as large as three feet high and seven feet in diameter in clearings of deciduous forests of the northeastern United States. The soil profile (a vertical section of the soil through all of its layers) at a mound is very different from soil profiles between mounds.

Various tropical species of termites erect huge mounds. Some of the hills are so large and plentiful that from a distance they resemble the huts of a human village. They can be 40 to 60 feet in diameter and up to 25 feet high. The transport of so much subsoil to surface hills has caused soil profiles over large areas of the tropics to be thoroughly reworked over long stretches of time.

Termite mounds can provide formidable obstacles for would-be cultivators of the land. Some of the structures are so tough they make sparks fly when hit with a hatchet. And if mounds are removed by bulldozing or dynamite, the termites can rebuild them with tremendous speed. Construction crews have removed mounds to clear airport runways, only to see takeoffs and landings imperiled within a day or two by newly erected mounds.

All Manner of Soils

Now that we have taken a look at what soils are made of (organic matter and the products of weathered rock) and the agents that create soils (the weather, plants, and animals), we will investigate some U.S. farmlands, both good and poor, to see how they got their characteristic features.[9] We will do this by considering the successes and problems of the early farmers in New England, the cotton belt, Mississippi River plantations, the Old Northwest Territory (now the north central United States), the Great Plains, and the arid West, while also exploring the origins of those soil properties that gratified or plagued the farmers. You should remember that the generalizations I make here apply to the most common soils in each area, but that the inherent variability of soils creates many exceptions in each location.

The stony, infertile soils so common in northern New England and eastern Canada defeated the best efforts of many farmers, whose abandoned fields reverted to forest. Now we know why those soils are so unproductive in their natural state: the soils have been intensely leached by acid waters. The strong acidity was created by the decomposition of a surface mat of litter from pines and other mostly coniferous species whose leaves contain relatively low amounts of such acid-neutralizing elements as calcium, magnesium, and potassium. The leaching has flushed many vital nutrients

9. See S. W. Buol, F. D. Hole, and R. J. McCraken, *Soil Genesis and Classification* (Ames: Iowa State University Press, 1973).

from the soil, and the strong acidity interferes with growth of crops not suited to such "sour" soils.

If you dig a pit in a well-developed example of those highly acid-leached soils you can see a striking view of distinct layers. The soil on top is porous, ashy gray, sandy material, washed clean of almost everything but quartz by the acid waters; below is a darker layer of soil where organic matter, iron, and aluminum have accumulated.

The acid-leached New England soils support good forest growth even though their agricultural uses are limited (if well fertilized, they can become productive, as in the potato-growing soils of nothern Maine). The forests were vital to the colonists, producing their soft-wood, shipmasts, and other structural timbers.

Soils in the cotton belt of the southeastern United States, while typically more fertile and less acid than those of New England, are generally relatively poor for agriculture in their natural state. The long growing season and the plentiful moisture allowed luxuriant plant growth at first, but after initial success many of the farmlands began to fail as soil nutrients became depleted and erosion removed topsoil. Large numbers of southern farmers—even some of the great plantation owners—had to move to fresher lands or, if unwilling to move, allow their tired lands to go fallow or put in subsistence crops to provide for the people and animals on the farm.

Why did so many southern soils begin to fail after a period of time? Because they are south of the advances of the most recent glaciations, many of the soil parent materials are old geologically, compared with the areas where glaciers brought fresh soil material. Intense weathering in the hot, humid cotton belt has created a high proportion of strongly weathered soils on these geologically old areas. As a result, southern planters were forced to apply large amounts of artificial fertilizers to their exhausted soils in order to retain their productivity. In 1919, for instance, Georgia, South Carolina, and North Carolina spent $147 million on fertilizers.

While southern cotton lands were generally old and relatively infertile, the rich floodplain along the Mississippi River was not. This region was described by Richard M. Ketchum in *The American Heritage Book of Great Historic Places:*

> What a man could see from these [Mississippi River steamboats] as he came within 200 miles of New Orleans was one of the great spectacles of the American scene. Within sight of each other, on both sides of the river and along the bayous, were the magnificent plantation houses—tall, pillared white residences built by the wealth derived from cotton and sugar. Cultivated fields of tassel-topped sugar cane and long, low rows of grayish-white cotton balls came down to the levee's edge, and all around was the lush foliage of the semi-tropics—clumps of live oak and cypress hung with Spanish moss, huge flowering bushes of camellias and magnolias, hedges of roses, and the sharp-pointed palmettos and Spanish dagger.[10]

Though drainage and protection from overflow were sometimes needed, the mild climate and rich soil were ideal for the cotton and sugarcane that were tended by slaves and transported to New Orleans by riverboat.

The soil's fertility was not surprising, since the floodplain included deposited sediments that had been

10. Richard M. Ketchum, ed., *The American Heritage Book of Great Historic Places* (New York: American Heritage Publishing, 1957), p. 232.

removed by erosion from some of the nation's best top-soils. These riverborne sediments were so nutritious for plants that they almost made good the claim in Mark Twain's *Life on the Mississippi* that "a man that drunk Mississippi [River] water could grow corn in his stomach if he wanted to."[11]

Another region that contained many soils more productive than typical New England or cotton-belt farms was the old Northwest Territory (now Indiana, Illinois, Iowa, Michigan, Wisconsin, and part of Minnesota). An important contributor of the fertile soils of this region was the glaciation of the ice ages. As the great glacial ice sheets slowly spread south, they gathered and brought with them huge amounts of soil and rock, which were left as fresh deposits when the glaciers subsequently receded during warmer conditions. In addition, fine materials were carried many miles below the leading edge of the ice sheets by streams flowing from melting glaciers. During the dry weather these unvegetated, unconsolidated fine sediments were very susceptible to wind erosion and created huge dust bowls of fine silty material called "loess." Loess deposits covered vast areas in the Midwest (southern and central Iowa and Illinois, southern Ohio and Indiana) and parts of the Northwest.

The fine, water-deposited sediments, windblown loess deposits, and soil parent materials are relatively young geologically, even though the last glaciers disappeared from northern Iowa and central New York about 12,000 years ago. Soils from geologically fresh materials are usually not drastically leached, and these young soils are generally richer in nutrients (and therefore crop-producing power) than older soils would have been.

11. Mark Twain, *Life on the Mississippi* (New York: New American Library, 1962), p. 29.

Some glacial soils are not so good, though. In the northern lake states and in New England, glaciers deposited parent materials that resulted in the sandier soils, which—particularly under coniferous forest—tend to be very acid, highly leached of nutrients, and relatively infertile.

The glaciers created an agricultural heartland in the central United States that is one of the most productive farming areas in the world. Settlers from the eastern seaboard and Europe poured into the territory, clearing forests, building log cabins and cottages, and preparing the lands for tillage. Soils under the cleared deciduous forests proved productive and relatively rich in nutrients compared with many New England or cotton-belt soils.

When the first European settlers began to move westward, it appeared to them that the vast sea of forest stretched endlessly before them. So to every farmer who came upon the open prairies, after leaving the dark woods, the sight was a wonder. "Bruised by the brushwood and exhausted by the extreme heat we almost despaired," reported one westbound settler as he approached the Illinois country in 1816. "A few steps more and a beautiful prairie opened to our view—lying in profound repose under the warm light of an afternoon's summer sun."[12]

Because settlers had not known farmland where trees did not grow, some avoided the grassy areas, fearing that they might be too sterile for crops. But in the 1830s and 1840s, after the development of a suitable plow for breaking the deep, thickly matted sod (see chapter 7), the prairies quickly turned into a marvelously productive farming area. In subsequent years set-

12. Paul Engle, "The Prairie and the Plains," in *The American Heritage Book of Natural Wonders,* ed. Alvin M. Josephy, Jr. (New York: American Heritage Publishing, 1963), p. 166.

tlers moved farther west onto the prairies of the Great Plains.

The richest soils in the United States, "the black earths," were formed in low-leaching areas of light rainfall on midwestern prairies. The breakdown of organic matter, especially the thick root systems of the grasses, created a black pigment that coated the soil to a depth of a foot or more. The reserves of organic matter and plant nutrients were so great that the soils produced years of bumper crops for the pioneer farmers without any additions of fertilizer.

During the 1870s and 1880s the rich black earth drew great numbers of land seekers from among immigrants, dissatisfied eastern farmers, and the ambitious young sons of large farm families. Farms of 10,000 acres were common, and there was even one of 100,000 acres. The farms were cultivated with the most modern equipment for that period; at times it was possible to see sweeps of 40 or 50 plows drawn by teams of sturdy workhorses moving abreast across a single farm.

For eight years before 1886 there had been so much rainfall on the Great Plains that responsible geologists predicted a permanent increase in the moisture of the climate. The richness of the soils had resulted from a drier climate with uncertain rainfall, however, and the inevitable return of that dry climate, in the form of a severe period of drought, brought the amazing farm prosperity to a disastrous end. Drought and frost combined to make crops almost total failures for all but two years between 1886 and 1895. Ruined settlers abandoned their farms in the Dakota Territory, Kansas, and Nebraska and returned in covered wagons, some bearing the legend:

IN GOD WE TRUSTED,
IN KANSAS WE BUSTED.

Thus was learned the lesson that the most fertile soils are not necessarily the most productive soils — unless they can be irrigated. The wetter prairie soils to the east, though more leached and less fertile, have a higher potential for production because of more dependable rain for the crops.

Soils in the arid and semiarid West, where the dry climate has reduced flushing of soluble byproducts of rock weathering through the soil, have also retained many indispensable plant nutrients; this is why irrigated valleys in the western states are among the best farmlands in the country.

Americans were shown the rich potential of such soils as a result of religious persecution. Brigham Young's Mormon followers, leaving the U.S. borders for the safety of what was then Mexican territory, were at first appalled by the desolation of their new site in the valley of the Great Salt Lake. By July 24, 1847, however, Young's small company of pioneers had watered, planted, and sowed upward of 100 acres with various kinds of seeds; nearly stockaded one public square, and erected one line of log cabins. There was near disaster from drought and crop-devouring crickets, but the colony overcame its problems; two years after its establishment, Forty-Niners on their way to California found a fertile and flourishing settlement with 5,000 inhabitants.

Toll of Time

28

3
Death, Decay, and Compost

Bodies That Didn't Decay

In May of 1950, archeologist Peter Glob was lecturing to his students at Aarhus University in Jutland, Denmark. He never finished the lecture because he was called away by a telephone message that would soon lead him to one of the most amazing examples of preservation of the past ever viewed by man. The superintendent of the police station at Sikeborg, 25 miles away, was calling to report that peat-cutters at nearby Tollund had discovered the body of a man in the peat. Many prehistoric artifacts had been dug up from the Danish peat bogs, so the superintendent of police had reason to contact the archeologist.

Professor Glob described what he saw that day:

> I started at once for Tollund Mose, a narrow peat bog among high, steep hills in a wild region of central Jutland.
>
> In the peat cut, nearly seven feet down, lay a human figure in a crouched position, still half buried. A foot and a shoulder protruded, perfectly preserved but dark brown in color like the surrounding peat, which had dyed the skin. Carefully we removed more peat, and a bowed head came into view.

As dusk fell we saw in the fading light a man take shape before us. He was curled up, with legs drawn under him and arms bent, resting on his side as if asleep. His eyes were peacefully shut; his brows were furrowed and his mouth showed a slightly irritated quirk as if he were not over-pleased by this unexpected disturbance of his rest.

That this rest had lasted 2,000 years was clearly shown by the seven feet of peat which had gradually formed above him throughout the centuries.

Now it was necessary to act swiftly to prevent the air from destroying this rare relic and to bring it as soon as possible under the care of a skilled conservator.

Careful hands covered the body again with peat, cut free the section in which it lay, and placed it in a wooden case. Thus the body was transported to the National Museum in Copenhagen.[1]

The ancient man from Tollund was naked except for a leather belt, a leather cap—and a braided leather rope with which he had been hanged.

Within a couple of years after this discovery, two other perfectly preserved bodies, also apparently victims of violent death, were uncovered by workmen cutting peat. A second victim, a 40-year-old man whose throat had been cut from ear to ear, was found six feet below the surface of a Jutland peat bog near Tollund. Fingerprints taken by the police department were as beautifully clear as those of a living person. Radioactive carbon dating showed that he had died about 300 A.D. The body of a 14-year-old girl, naked except for an ox-skin collar, was freed from peat in northern Germany.

1. See Geoffrey Bibby, *The Testimony of the Spade* (New York: Knopf, 1956), pp. 397–401, for Glob's report and a description of other bodies recovered from bogs.

The exact cause of death was unclear, but a woven band still blindfolded the girl's eyes. Identification and counting of ancient pollen grains preserved beside the body showed that she had died about 2,000 years before. Changes in content of windborne pollen parallel the trend of changes in vegetation, and where the sequence of layers of pollen has been carefully studied, as in Scandinavian peat bogs, the pollen compositions can be correlated with periods in the past.

The three finds caused a worldwide sensation, and hundreds of thousands of people have flocked to a museum to see these Europeans of a hundred generations ago. Their violent deaths, which occurred at approximately the same time, present a fascinating murder mystery. Tacitus, the Roman historian, described a ceremony dedicated to the fertility goddess in which some of the attendants were killed and thrown in a lake. This could be the explanation for the bodies in the bogs. Tacitus also wrote that throughout the violent Germanic area, traitors were hanged, and cowards and perverts drowned in swamps.

Another fascinating question concerns the means by which ancient bodies could be so perfectly preserved that even fingerprints and facial expressions are as clear as those of a living human. Preservation over very long stretches of time requires conditions that prevent the growth of decay microbes. In bogs, water acts as a partial preservative by shutting out the oxygen needed by most decay microbes; when the bogs are in cold climates, as are those that contained the three ancient bodies, low temperatures exclude many species of decay organisms and slow down the activity of others. In addition, bog water in which peat has soaked is rich in tannic acid, a remarkable preservative. Tannic acid from vegetable sources is the old-fashioned substance that was used in tanning animal skins to make them resistant to decomposition and turn them into leather.

Bodies in frozen areas remain virtually free of decay microbes. In 1982 the frozen bodies of a family of five Inupiat Eskimos were found in their ice-covered home on a bluff overlooking the Arctic Ocean near Barrow, Alaska.[2] It appeared that strong winds had pushed a thick cover of snow over the sod home, trapping the doomed family. Measurements based on carbon dating showed that the almost perfectly preserved Eskimos had died around 1510. Autopsies performed on two of them—a woman about 42 years old and another in her 20s—demonstrated that the bodies had changed so little since death that the health problems of 470 years earlier were still clearly visible. The lungs of both women were blackened by smoke from the poorly ventilated home. The older woman had hardening of the arteries as well as heart damage, apparently from a serious infection when she was young. In addition, she had trichinosis (a disease caused by parasitic worms), probably the result of eating inadequately cooked polar-bear meat. Both women had osteoporosis, a condition producing fragile bones, which may have been caused by the lack of sunlight during the long arctic winter.

Other ancient bodies, though not as perfectly preserved as those in bogs or ice, have been discovered in very dry locations, such as Egypt. Too little moisture, as well as too much, can limit the decay microbe's action. Before the ancient Egyptians began to embalm their dead and inter them in structures, unembalmed bodies were buried in shallow pits, where the dry soil preserved some so well that they still have hair and skin.

2. Michael R. Zimmerman and A. C. Aufderheide, "The Frozen Family of Utquiaquik: The Autopsy Findings," *Arctic Anthropology* 21, no. 1 (1984): 53–64.

Decay

Why is such a lack of decay so very rare? In most situations with sufficient moisture and oxygen and favorable temperatures, hordes of decay microbes thrive and soon begin their work. If you note the changes in rotting organic matter—from fresh tissues to a dark, crumbly, unrecognizable mass—they all appear to be caused by some mysterious invisible agent. A powerful microscope gives the explanation: dead matter, whether it was once Julius Caesar or fallen leaves, is riddled with countless microscopic spherical, spiral, rod- or filament-shaped bacteria and threadlike fungi. These microscopic plant forms are digesting the organic matter for their food just as surely as would the stomach of an animal. The microbes break down not only natural organic substances but also synthetic pesticides. Most pesticides have a relatively short half-life (the time it takes for half the active substance to be destroyed or altered) of a few weeks in the soil. The microbes have no mouths, but soluble foodstuffs diffuse into their cells through the cell walls. These microbes also release enzymes to change insoluble foods into soluble forms.

The invisible army of bacterial and fungal decay microbes is doing incredibly important work: retrieving indispensable substances from the dead for the living. If they did not, so many vital elements would eventually become permanently tied up that disaster would inevitably follow for all life.

If there were no decay microbes, the supply of carbon—the most important element in living things—could rapidly be depleted. The ultimate source of carbon is the carbon dioxide that makes up only one part in 300 of the gases in the earth's atmosphere. Carbon dioxide gas is fixed into organic compounds by chlorophyll-containing plants. When the decay fungi and bacteria decompose dead material, they complete the cycle by releasing carbon dioxide into the air. The microbes also release such essential elements as nitrogen, phosphorus, and sulfur. If organic matter were not broken down to release these life-essential substances, living things would meet a catastrophic end, and the accumulated dead but undecomposed material would be slowly turned into coal during the following eons.

The vital services performed by the decay microbes also convert the dead materials into the vast, heterogeneous collection of carbonaceous compounds known as humus. Humus, which holds large amounts of water and readily available plant nutrients, plays an almost miraculous role in conditioning soils—particularly those heavy in clay—by making them loose and open, well drained, and well aerated. Humus has this effect because the organic materials aggregate soil particles into larger units, resulting in the granulation of soil particles into clumps separated from other clumps by spaces that allow easy penetration of water, air, and roots.

But the microbes are not working out of altruism. They have to break down dead matter for their source of energy and to build up their own tissues. The decay microbes, like entrepreneurs, are to be found wherever a profit is to be made—an energy profit. The dull-colored, darkness-inhabiting microbes are not able to make their own food, unlike the green chlorophyll-containing plants that live in the sunlight and manufacture carbohydrate foods by using the sun's energy to combine carbon dioxide and water. The potential energy in plant carbohydrates sustains not only the vegetation that can manufacture this food but all animal life *and* the decay microbes. When those hungry microbes break down the food in animal or plant remains by a series of complicated chemical reactions, they tap the energy and use it to build and maintain their own cells.

Each of the microbes is extremely minute, but their total is astounding in size and effect. The weight of this busy army in a grassland, for instance, is far greater than that of all large mammals, rabbits, mice, gophers, toads, snakes, birds, grasshoppers, spiders, and other types of animal life in a comparable area. If microbes had minds and could think, they would probably conclude that they are the most important form of life, with all other forms existing only for their benefit as food.

When a rabbit dies or when leaves fall, so that more dead matter appears in the soil where there was little before, the numbers of microbes in that soil will increase enormously in a very short time. The microbes first attack the substances that decompose most readily, such as sugars and cellulose. When the easily digested materials are used up, most of the microbes die, but the remaining ones attack the more resistant lignin, waxes, fats, or even the soil humus. As microbes die, they themselves may constitute up to half the total soil organic matter. The staggering numbers of dead microbes, the greatest mass starvations on earth, are decomposed and consumed by other microbes—and so the never-ending cycle goes on. Nature is indeed efficient.

The microbes do not always carry out their work in complete harmony. One mechanism that has brought enormous benefit to humans is the production of antibiotics by some microbes in order to kill or retard others and thereby gain more nutrients for themselves by eliminating competition. Many important antibiotics, including penicillin and streptomycin, have been obtained from soil microbes.[3]

3. Soil microbes and decay are covered in great detail in Martin Alexander, *Introduction to Soil Microbiology*, 2nd ed. (New York: Wiley, 1977).

Compost

An interesting and practical example of the decomposition of organic matter can be seen by making a compost pile and watching it decay. Nearly any natural organic substance can be used: grass clippings, kitchen garbage, leaves, straw, cornstalks, winery refuse, and the like. These residues should be cut less than six inches in length, and the heaps should be no more than six feet high so that air can enter the bottom of the pile. The pile should be kept as wet as a squeezed-out sponge, with water and nitrogen fertilizer (a quarter pound of nitrogen per cubic foot of material) applied to the layers as the heap is built up.

The compost should be turned every three or four days to expose pockets of poor aeration to air. This causes faster decomposition and a rapid evolution of heat. The compost is finished when it cools, has a dark color, and is crumbly and odorless. The process may take as little as three to four weeks under favorable conditions.

An interesting aspect of compost piles is the high temperature generated when the decay microbes release heat in rotting the dead matter. If you build a compost heap during warm weather, you will probably see heat waves rising from it within a day or two, and the temperature within may be greater than 150 degrees Fahrenheit. This elevated temperature destroys weed seeds and many parasitic organisms in the plant and animal matter.

Because many decay microbes can't live at the high temperatures found in compost piles, decomposition under such conditions is achieved primarily by filamentous bacteria called actinomycetes. Though most actinomycetes do best in temperatures between 75 and 80 degrees, some species thrive in much greater heat; in fact, they become so numerous that they often color compost the white or gray of their own bodies. There may be as many as ten billion heat-loving actinomycetes per gram (it takes 454 grams to make a pound) at 120 to 150 degrees.

It may seem surprising that no advantage appears to be gained by rotting materials in the compost heap rather than letting them rot in the soil to add plant nutrients and increase soil humus. The following facts should be considered:

> Composting is a convenient way to store waste organic materials if it is impossible or impractical to add them to soil as they become available.

> Though unrotted material holds little water and is fibrous and coarse, adding it directly to heavy clay soils may be advantageous because it can make the soil more open and free draining. In contrast, well-rotted compost may best reduce the openness of excessively free-draining sandy soils.

> Allowing material to rot in the soil will often add more to the nutrient and humus content of the soil than if it is first composted. In experiments conducted for years at Rothamsted in England (a famous agricultural experiment station with a very long history: see chapter 4), applications of straw and nitrogen plowed into the soil were compared with the same amount of straw and nitrogen first composted and then applied. The fields with plowed-in straw and nitrogen had better crop yields than the fields where the composted material was added—probably because nitrogen and potassium were lost during the composting process. Much of the nitrogen can escape as ammonia gas and the potassium by leaching.[4]

4. Rothamsted Experimental Station Report for 1951, pp. 135–40.

There is no question that compost is an excellent soil conditioner, but it may have limitations as fertilizer, as we will see in the next chapter.[5]

Toxic Organic Matter

Indiscriminate use of any and all plant material as soil additions could lead to serious problems. Some plants produce substances that are poisonous to other plants (black walnuts, for example, produce a toxic substance called juglone), so not all species are desirable as soil conditioners.

Probably the most intriguing method by which plants suppress competitors is through production of toxic organic chemicals that inhibit the growth of other plants — a process called *allelopathy*.[6] In the southwestern United States, for example, some low, fragrant chaparral shrubs growing on clay soils have a bare strip a yard or two wide separating them from adjacent annual grassland. C. H. Muller, a botanist at the University of California at Santa Barbara, speculated that the bare strip is caused by the same substances that make the shrubs so deliciously fragrant. Muller concluded that some of those aromatic vapors, especially cineole (which is also the source of fragrance in eucalyptus) and camphor, become attached to the surfaces of the clay soil particles, where they suppress the germination and growth of annual plants during the rainy season.

Another example of plant exclusion that appears to

5. Directions for making compost may be found in Ortho Books, *All about Fertilizers, Soils and Water* (San Francisco: Chevron Chemical Company, 1979), pp. 46–47. Some of the limitations of compost are discussed in E. W. Russell, *Soil Conditions and Plant Growth*, 10th ed. (London: Longman, 1973), pp. 271–73.

6. A good reference about allelopathy was written by Elroy L. Rice: *Allelopathy* (New York: Academic Press, 1984).

be caused by allelopathy is the restriction of Norway spruce from heather-dominated heaths in Great Britain. The roots of heather produce a substance that suppresses the growth of fungi associated with the spruce tree's roots. These fungi, called mycorrhizae, are apparently vital to the spruce, as they absorb large amounts of nutrients that can then be used by their host tree.

Are plants that give off allelopathic substances heartless enough to prevent the growth of their own offspring? In some instances indeed they may be: silk oak (*Grevillea robusta*) seedlings are killed under the parent trees along with any others present. Next in line are certain suicidal plants which, over a period of time, produce chemical effects in the soil that destroy their own populations. They are replaced by other species more tolerant to the chemicals.

Though many plants produce toxic chemicals that can inhibit the growth of other plants, the presence of toxin in one plant does not mean that it will be released into the environment of another plant in sufficient quantities to have a fatal effect. While allelopathy seems to be important in determining vegetation composition in some situations (scientists are studying the possible use of allelopathic crop plants for weed control), its overall importance in nature is not yet clearly understood.

Buggy Ride

4
Muck, Mystery, and the Good Old Days

A Chronicle of Fertilizers

An ancient saying about plant growth is that "corruption is the mother of vegetation." This does not mean that one must be a wicked person to be a successful gardener or farmer but instead refers to early observations that decaying manures, composts, and dead animals (or parts of animals, such as blood) all increased soil fertility. Despite this early insight, the question of what substances plants require for growth was long the subject of acrid debate.

The first chemical experiments to investigate plant nutrition were carried out by Jan Baptista van Helmont (1577–1644), a Flemish chemist. Van Helmont grew a potted willow tree in 200 pounds of soil, adding only water, and found that after five years the tree had gained 164 pounds, while the soil had lost only two ounces. He concluded that water was the sole nutrient of plants because "the 164 pounds of wood, bark, and root arose from the water alone."[1] He thought that the apparent loss of two ounces of soil was probably a mere error in measurement.

1. J. B. van Helmont, *Ortus Medicinae Id est, Initia Physica in audita, Progressus Medicinae novus, in Morborum Ultionem, ad*

We know now that most fresh plant tissue, generally anywhere from 94 to 99.5 percent, is made of carbon, oxygen, and hydrogen, which come mostly from the air and from water; only the small remainder of the plant tissue is made from soil ingredients. This is what caused van Helmont's confusion. Even though plants are made *mostly* from air and water and only a small part from soil, they will not grow properly if the nutrients in soil are insufficient. It was ironic that van Helmont did not take the air into account, for he was the first to study airlike substances. He invented the word "gas," and a gas that he discovered and named *spiritus sylvestris* (spirit of the wood) turned out to be carbon dioxide, the major source of a plant's subsistence.

Other primitive theories were equally wrong. One conjecture was that plants get everything they need from air. Another was that plants take up organic matter directly as the source of all their nutrients; in chapter 1 we looked at Jethro Tull's idea that plants feed on minute particles of soil. Tull thought that the soil particles should be made fine by cultivation to aid in feeding.[2]

While the various early theories about the causes of fertility were being concocted, some close to the truth and some plainly mistaken, the spreading of natural organic materials remained the primary means of fertilization. The Europeans seem to have appreciated the value of such materials more than the early American settlers, who often allowed animal manures to pile up

Vitam Longam (Amsterdam: apud Ludovicum Elzevirium, 1652), p. 88.

2. The early investigations of plant nutrition are summarized in E. W. Russell, *Soil Conditions and Plant Growth*, 10th ed. (London: Longman, 1973), pp. 1–16. See also F. G. Viets, Jr., "A Perspective on Two Centuries of Progress in Soil Fertility and Plant Nutrition," *Soil Science Society of America Journal* 41, no. 2 (March–April, 1977): 242–49.

rather than spreading them on the fields. The frequent American practice of clearing land, allowing it to "wear out," and then moving to fresher soils was more like the "shifting cultivation" of the wooded tropics (to be discussed in chapter 11) than like any modern counterpart.

Some farmers were not so careless of their soil's fertility, however. The great demand for produce in the developing New York vegetable market near the end of the eighteenth century inspired farmers in that area to search furiously for fertilizer materials. Their efforts were described by the Reverend Timothy Dwight in his *Travels in New England and New York*:

> The inhabitants, with a laudable spirit of enterprise, have set themselves to collect manure, wherever it could be obtained. Not content with what they could make, and find, on their own farms, and shores, they have sent their vessels up the Hudson, and loaded them with the residuum of potash manufactories; gleaned the streets of New York [of street sweepings]; and have imported various kinds of manure from New Haven, New London, and even from Hartford. In addition to all this, they have swept the sound; and covered their fields with the immense shoals of whitefish with which in the beginning of summer its waters are replenished. No manure is so cheap as this, where the fish abound; none is so rich; and few are so lasting.[3]

The whitefish, caught in huge quantities in seine nets during June and July, were sold for a dollar a thousand, and about 10,000 fish were used to fertilize an acre. They were frequently laid in furrows and covered

3. Timothy Dwight, *Travels in New England and New York*, vol. 3 (New Haven, Conn.: Timothy Dwight, 1822), p. 303.

with plowed soil; sometimes they were placed singly on hills of corn and hoed in; at other times they were combined with additional organic debris to form compost, then carted into the fields and spread in the same manner as manure from the stable. The odor of those farms must have been overwhelming!

The fertilizer practices around New York in "the good old days" would have gladdened the hearts (if not the noses) of many present-day organic farming enthusiasts, but the old theories of plant nutrition were heading for their downfall, at least in the scientific community. In 1840, Justus Baron von Liebig, a renowned German chemist, attacked the old humus ideas with ferocious scorn in a paper presented at an important scientific meeting. Liebig thought that manures did not provide anything magical for plants but merely restored some essential minerals to the soil; therefore, why not add the minerals themselves—pure, clean, and odorless—and eliminate the need to handle the stinking manures? Liebig's pioneering work on artificial fertilizers made him the father of the commercial fertilizer industry.

Liebig also ridiculed the idea that plants derive their carbon from humus rather than from the carbon dioxide of the air. That idea was held by many plant physiologists of the day, despite much evidence to the contrary. Liebig wrote that the experiments cited by adherents of the humus theory were all "valueless for the decision of any questions." He stated that "the experiments are considered by them as convincing proofs, whilst they are only fitted to awaken pity."[4] Liebig was a towering figure in chemistry—one of the primary influences in making chemistry almost a German monopoly

4. Justus Liebig, *Chemistry in Its Application to Agriculture and Physiology*, ed. Lyon Playfair, 2nd ed. (Cambridge: John Owen, 1842), pp. 61–62.

in the nineteenth century—and his paper effectively killed the humus theory as a believable explanation. Few scientists have dared since that time to suggest that carbon in plants comes from any source other than the carbon dioxide of the air.

In 1843, soon after Liebig presented his paper, an agricultural experiment station was established in Rothamsted, England. Farmers were slow to believe that chemical fertilizers could do more than stimulate a crop and thought that their use would ultimately exhaust the soil. But plots at Rothamsted, fertilized with chemicals and sown always with the same crops, showed that good yields could be produced even after a century of such treatment. This demonstrated that soil fertility could be maintained by use of chemical fertilizers alone; if soils become depleted despite the use of chemical fertilizers, it is because of insufficient amounts of these substances and improper soil management.

How many elements are indispensable for the growth of plants? Essentially, all plants require only 16 of the 92 naturally occurring elements. Taken from *air and water* are (1) carbon, (2) hydrogen, and (3) oxygen. Taken from *soil* and needed in relatively large amounts are (4) nitrogen, (5) phosphorus, (6) potassium, (7) calcium, (8) magnesium, and (9) sulfur; needed in smaller amounts are (10) iron, (11) manganese, (12) boron, (13) molybdenum, (14) copper, (15) zinc, and (16) chlorine.

Regardless of the original source of the soil nutrients, whether natural organic matter or inorganic chemicals, fertilizer must break down into simple forms of chemicals that can be absorbed from the soil water by roots. The essential elements are all absorbed from soil as ions—atoms or groups of atoms (depending on the specific element) with an electrical charge. The ions absorbed by plants are identical, whether from an organic or an inorganic source. Experiments at various state and federal experiment stations have shown no difference in the vitamin or mineral content of crops grown with organic as compared with inorganic nutrient sources. Nor is there evidence that adding vitamins or plant hormones to the soil improves crop production or vitamin content.

The fact that plants can be grown in solution cultures containing only mineral ions, without differing in yield or quality from plants grown in fertile soils, also strongly supports the theory of mineral nutrition and undermines the theories on which organic farming is based. Although hydroponics (as such a system is called) can be successful, it is not easy. Hydroponic farming is more expensive and inconvenient than growing plants in soil. Because plants cause the nutrient solutions to change, it may be necessary to replace the solutions every few days or even daily. Most plants grow far better if oxygen is provided for roots by forcing air through the solution; in addition, the solution must be kept dark to prevent the growth of algae, and there may be some minor problems in keeping the plants in an upright position. Frequently, the plants are grown in containers of pea gravel or coarse sand to support the roots, while the solution is trickled or flushed through periodically to provide water and nutrients.

Hydroponics has been used extensively for scientific research and commercial vegetable production. Nutrient levels can be carefully controlled and their effects studied in plant experiments. During World War II, food was hydroponically grown for military forces in remote locations where weather or soil conditions did not allow normal farming.[5]

5. Howard M. Resh, *Hydroponic Food Production*, 2nd ed. (Santa Barbara, Calif.: Woodbridge Press, 1981).

Muck and Mystery

The success of humus-free hydroponic solutions and all the other accumulated evidence against organic matter as a source of superior nutrients have not discouraged the organic theorists. In recent decades we have seen a new humus theory gain popularity among organic gardeners and farmers. The organic movement has been jokingly called "muck and mystery" because some of its enthusiasts have practiced it as though it were a religion—the "mystery" having even greater importance than the "muck." They believe that natural organic matter has some mystical life-giving properties that can never be replaced by inorganic fertilizers.

Sir Albert Howard (1873–1947), a British colonist in India, is venerated as the founder of the "organic" movement. Howard, researcher at the 300-acre farm of the Indore Institute of Plant Industry in India, developed large-scale composting techniques based on Indian and Chinese folk methods. He returned to England in 1931, after 30 years in India, and popularized his ideas on composting.

Sir Albert believed that the forest floor was the perfect example of a natural standard of soil fertility. While always extolling the virtues of natural organic matter, he changed his ideas several times about the value of synthetic fertilizers. In a 1916 lecture he argued that chemical fertilizers were a waste of money; he maintained that organic matter alone, and the good aeration it promoted, would allow microbes to provide sufficient amounts of nutrients to feed the world. Fifteen years later he had reached a very different conclusion. In a book written in 1931, *The Waste Products of Agriculture: Their Utilization as Humus,* Sir Albert stated: "The full possibilities of humus will appear when the dressings of compost are supplemented by the addition of suitable artificials [chemical fertilizers]. The combination of the two, applied at the right moment and in proper proportions, will open the door to the intensive crop production of the future. Humus and artificials will supplement one another."[6] Shifting his view again in his later writings, Howard attacked the common use of synthetic fertilizers as one of the world's "greatest calamities" and referred to hydroponics as "science gone mad."[7] Why did Sir Albert Howard, after concluding in his final summary of his work in India that artificial fertilizers should be used to reach maximum yields, then condemn their use so fiercely? In *The Soil and Health,* first published in 1947 (the year of his death), he asserted that the use of synthetic chemical fertilizers leads to imperfectly synthesized protein in leaves and results in many of the diseases found in plants, animals, and human beings: "When these proteins are manufactured from freshly prepared humus and its derivatives, all goes well; the plant resists disease and the variety is, to all intents and purposes, eternal. But the moment we introduce a substitute phase in the nitrogen cycle by means of artificial manures, like sulphate of ammonia, trouble begins which invariably ends with some outbreak of disease, and by the running out of the variety."[8]

Howard was convinced that he had found the nature of disease in the writings of J. E. R. McDonagh, whom he called "a distinguished investigator of human diseases."[9] McDonagh believed that all proteins alter-

6. Albert Howard and Yeshwant D. Wad, *The Waste Products of Agriculture: Their Utilization as Humus* (Oxford: Oxford University Press, 1931), p. 112.

7. Albert Howard, *The Soil and Health* (New York: Schocken Books, 1972), pp. 75, 194.

8. Ibid., p. 189.

9. Ibid., p. 187.

nately expand and contract as a result of changes in weather:

> If the sap in plants does not obtain from the soil the quality of nourishment it requires, the protein over-expands. The over-expansion renders the action of climate an invader. That is to say climate, instead of regulating the pulsation, adds to the expansion. The over-expansion results in a portion of the protein being broken off, and this broken off piece is a virus. The virus, therefore, is formed within, and does not come from without, but protein damaged in one plant can carry on the damage if conveyed to other plants. The protein in the blood of animals and man suffers the same damage if it fails to obtain the quality food it needs.[10]

McDonagh's theory of disease is nothing but moonshine, but Howard used it to condemn synthetic fertilizers and extol the health-giving virtues of natural organic matter: "If, therefore, we see to it in our farming and gardening that the effective circulation of protein from soil to plant, and then to livestock and mankind is maintained, we shall prevent most of the departures from health—that is to say disease—except those due to accidents or to abnormal climatic conditions."[11]

In America, the late J. I. Rodale, an early reader and follower of Howard, promoted the organic farming and gardening movement that has acquired many zealous adherents in the United States. Books and magazines about organic growing are published by the Rodale Press, and organic methods are studied at the Rodale experimental farms.

Natural organic fertilizers are prized by the organic

10. Quoted in ibid., pp. 187–88.
11. Ibid., p. 188.

adherents because they improve the soil's physical condition and because they release plant nutrients slowly and reduce the danger of overfertilizing. All schools of thought readily admit the value of organic matter in improving and conditioning the soil, but the alleged superiority of natural organic fertilizers over synthetic chemical fertilizers does not hold up under scrutiny.

Beautiful gardens can indeed by grown using only organic materials as a nutrient source. However, the high yields of modern agriculture require the concentrated, easily available forms of nutrients in synthetic fertilizers because the major plant nutrients are not present in organic matter in sufficient concentrations to sustain maximum crop growth.

It is much easier to duplicate the forest floor—the ideal of Sir Albert Howard—in small garden plots, which can be heavily supplied with waste organic matter, than on farmlands. A substantial decrease in soil organic matter content occurs when a virgin soil developed under forest or prairie is brought under cultivation. In nature all the organic matter is returned to soil, but under cultivation much of the organic matter is removed at harvest. Moreover, soil tillage breaks up the organic residues and exposes them to easier attack by microbes, thereby increasing the loss of carbon as it escapes in the form of carbon dioxide gas. The rate at which organic matter decays increases very rapidly as the organic matter content is raised. So it would be very difficult and expensive to maintain organic content at as high a level as occurs in soil in the virgin state. Farmers can't be expected to hold organic matter at higher levels than are consistent with crop yields that pay best. Even the most productive farm will likely have considerably less organic residue than nearby soil in an undisturbed location.

If added organic residues are very low in nitrogen

content, as are sawdust and grainstraw, they can temporarily rob crops of their nutrients. The soil microbes, like vegetation, require nutrient elements. While breaking down organic matter, the microbes take the nutrients they need from the soil, and if the nutrient supply is insufficient to provide for both microbes and plants, plant growth will slow down. The addition of chemical fertilizers can overcome this problem.

Organic materials that do a much better job of supplying nutrients in excess of microbial needs are those higher in nitrogen: animal manures, legumes, grasses, and mustards. At the present time, organic matter in the form of animal manures and sewage sludge is being considered as a possible source of agricultural nitrogen, given the increasing expense of commercial nitrogen fertilizers and the need to dispose of wastes from feedlots and other sources.

Commercial Fertilizers—Where They Come From

How are the nutrients most often supplied in the largest amounts by commercial fertilizers (nitrogen, phosphorus, and potassium) acquired in such enormous quantities for artificial fertilizers?[12] Phosphorus and potassium are mined as ores. Phosphate rock deposits are plentiful in North Africa, North America, and the Soviet Union. Rock phosphate is treated with sulfuric acid to change it into a more soluble form for fertilizer. Potassium is found throughout the world in both soluble and insoluble forms. Today only the soluble forms are economically feasible as fertilizer.

One discovery that had a great effect in increasing crop yields for the good of humanity was the result,

12. D. A. Russel and G. G. Williams, "History of Chemical Fertilizer Development," *Soil Science Society of America Journal* 41, no. 2 (March-April, 1977): 260–65.

ironically, of the pre–World War I arms race. At that time, most of the nitrogen for the world's explosives came from enormous nitrate deposits in a barren Chilean desert, and German military planners foresaw disaster should their supply of nitrates be cut off by a wartime blockade. This chilling vision was the chief incentive for German scientists to invent a process, in 1910, by which the nitrogen gas that makes up 80 percent of earth's atmosphere could be fixed into forms usable for explosives. Nitrogen and hydrogen gas are passed over a metal catalyst at 900 degrees Fahrenheit and a pressure several hundred times greater than the earth's atmospheric pressure at sea level. This procedure makes ammonia, which in turn can be converted to other forms of nitrogen — for fertilizers as well as explosives. Germany might well have collapsed before 1918 for lack of explosives if it had not been for the nitrogen fixation industries built between 1914 and 1916.

When the German scientists learned how to capture atmospheric nitrogen in useful forms, they accomplished what some very important species of microbes routinely do. The nitrogen-fixing microbes can be classified broadly in two groups: (1) bacteria that live in root nodules of several dozen species of plants (mostly legumes: alfalfa, clover, peas, etc.); and (2) "free-living" bacteria (not confined to root nodules), blue-green algae, and even some fungi.

Because atmospheric nitrogen, the ultimate source of this vital element for all life, cannot be used directly by the great majority of species on earth, there has been a very long period of dependence by most living things on these comparatively few species of microbial nitrogen fixers for much of the nitrogen needed as essential parts of proteins, protoplasm, chlorophyll, nucleic acids, and enzymes. Cycads and ginkos, primitive plants of which species still exist, can be traced back some 300 million years. Their roots contain bacteria that have the ability to fix nitrogen. A small but important amount of nitrogen is changed into usable forms by phenomena such as lightning, cosmic radiation, and meteor trails.

Now, with the advent of modern agriculture, the amount of nitrogen created annually as fertilizer by industrial fixation and by large-scale cultivation of nitrogen-fixing legumes probably exceeds the amount of nitrogen that was made available each year to living things by microbes and natural phenomena on all the earth's surface before the beginnings of agriculture. All this additional nitrogen has played a major role in the large increases in crop yields achieved during the past few decades.

Explosive Fertilizer

Because identical processes are used to fix nitrogen (make it usable) for fertilizers and explosives, you may ask whether any fertilizers are explosive. One nitrogen-containing fertilizer in common use, ammonium nitrate, can be highly explosive if not stored and handled properly. If it becomes contaminated with fuel oil or other combustible materials, it can explode; in fact, it is often mixed with oil and used for blasting.

On the morning of April 16, 1947, the freighter *Grandcamp*, loaded with ammonium nitrate fertilizer, caught fire and exploded with incredible violence at the port city of Texas City, Texas. The blast flooded docks and streets with a surge of water, capsized small boats in the harbor, and demolished several warehouses. A nearby steel barge was flung from the water, ending up 100 yards inland. Two light airplanes were obliterated in the air, and people were blown through doorways a mile away in the business district. The explosion was heard 150 miles away, registered on a seismograph in Denver, and killed many firefighters and spectators who had gathered on the docks. The Monsanto chemical plant, 700 feet from where the *Grandcamp* was docked, blew up a few minutes later. That blast demolished much of the business district, and set fires along the two-mile waterfront and through the rest of the city. At 1:11 the following morning the freighter *High Flyer*, also filled with ammonium nitrate, exploded in the harbor.

Texas City was virtually destroyed by the chain reaction of explosions caused by the detonated fertilizer. Property damage estimates totaled more than $125 million. At least 460 people died, but because the dock area contained many migrant workers without permanent addresses or known relatives, some estimates put the final toll as high as 1,000 dead and 3,000 injured. "In four years of war coverage," wrote Associated Press correspondent Hal Boyle, "I have seen no concentrated devastation so utter, except Nagasaki."[13]

One of the main reasons for the importance of nitrogen in living organisms and explosives is that many of the chemical compounds in which nitrogen is linked to other types of atoms absorb considerable energy in their formation. This energy may be released slowly, as in the breakdown of protein in your body, or violently, as in the decomposition of TNT.

Ammonium nitrate poses no danger, with a few precautions (no smoking nearby; no storage near steam pipes or wiring; keeping away from combustible materials such as gasoline, oils, paints, sulfur, straw, or shavings; and throwing out the empty bags—they are flammable). Any of the many other non-explosive forms of synthetic nitrogen fertilizer may also be used.

13. Hal Boyle, quoted in Irving Wallace, David Wallechinsky, and Amy Wallace, "Significa," *Parade* (January 23, 1983): 16.

The Organic Grower vs. the Agribusiness Enterprise

We have reviewed the development of both natural and synthetic fertilizers, and now I will try to compare the practicality of depending only on organic materials for fertilizer with that of using artificial fertilizers. Let us imagine two adjacent hypothetical farms: one run by agribusiness and the other by an organic grower. Let us further suppose that the agribusiness enterprise and the organic grower both have new computers and calculate the amounts of fertilizers they must respectively apply to wind up with the same amounts of nutrients per acre per year.

The agribusiness wants 200 pounds of plant-available nitrogen per acre per year, so it will apply slightly more than one-fifth of a ton per acre of synthetic urea fertilizer (46 percent nitrogen content) annually.

The organic grower also wants 200 pounds of nitrogen per acre per year, but much greater amounts of natural organic fertilizer must be applied.[14] Using dry corral manure with a nitrogen content of 1.5 percent, the organic grower then calculates how much manure to apply each year to keep levels of plant-available nitrogen at 200 pounds per acre each year (because it takes years for all the nitrogen in the manure to be converted to plant-available forms by decay microbes, the quantity of manure that must be added in later years can be reduced). The first year, the organic grower will have to spread 19 tons of manure per acre—88 times more weight of fertilizer than the agribusiness needed to get the same amount of available nutrients. More-

14. The following estimates were determined from P. F. Pratt, F. E. Broadbent, and J. P. Martin, "Using Organic Wastes as Nitrogen Fertilizers," *California Agriculture* 27, no. 6 (June, 1973): 10–12. See also Council for Agricultural Science and Technology, *Organic and Chemical Farming Compared*, report no. 84 (Ames, Iowa: CAST, 1980).

over, the organic grower is hauling more than 1,300 cubic feet of manure per acre to get that 19 tons, while the agribusiness would need a little more than 9 cubic feet of urea—much less than 1 percent of the bulk handled by the organic grower, who would probably feel the pressure of hauling and spreading fertilizer if the farm was of any size. And if the hauling were contracted out, the cost—even if the manure were obtained free of charge—would likely substantially exceed the cost of fertilizing the agribusiness farm.

In the second, third, fourth, and fifth years of fertilizing the organic farm, 14, 12, 12, and 11 tons, respectively, of manure per acre would have to be applied. By the tenth year that would drop to 9 tons per acre, and during the fifteenth to twentieth years only 8 tons per acre would be needed. But even these figures greatly exceed the weight and bulk of urea applied to the agribusiness farm.

Would the organic grower at least have better soil because of the favorable effects of organic matter on its physical conditions? The answer is maybe and maybe not. The bumper crops grown on the agribusiness farm could leave a lot of plant residues, especially root systems, so the organic matter there would be far from depleted. Also, the large amounts of manure on the organic farm could lead to substantial salt accumulations. In some moderately saline soils, until the salts leached out, yields would probably be reduced.

If the organic grower decided not to haul all that manure, the 200 pounds per acre per year of nitrogen could still be added by growing nitrogen-fixing legumes. Beans or peanuts fix only about 40 pounds per acre per year, so they will probably remove nitrogen from the soil rather than adding it. Red clover might fix 100 to 150 pounds per acre—still too low. A good crop of alfalfa can fix 200 to 250 pounds per acre per year, but not all of it becomes available right away; the nitrogen will be released over a number of years by the decay microbes. And if the alfalfa is harvested, a considerable amount of the fixed nitrogen will be removed in the tops. The nitrogen from a legume with high-fixing capacity like alfalfa can be used to support growth of non-legumes in a crop rotation system, and may be a desirable way of reducing disease buildup and erosion (the legume often provides greater ground cover and protection from erosion than a non-legume like corn). However, total yields of non-legumes would be much higher using continuous cropping and synthetic fertilizers. If the organic farm were in certain parts of New Zealand where clovers grow year round and may fix 500 pounds of nitrogen per acre per year, non-legume crops in rotation could be well supplied with nitrogen for a full growing season by the legumes alone.

The organic farm's nitrogen supply problems will not be the only difficulty in keeping up with the nutrients applied to the agribusiness farm. If phosphorus and/or potassium are deficient, these nutrients will be applied in soluble forms of chemical fertilizers.

The organic grower will discover that phosphorus is present in organic matter in even smaller quantities than nitrogen and becomes available slowly with decay. Organic farming literature recommends the use of rock phosphate, which dissolves too slowly under the best of conditions to do much good for short-season crops and which hardly dissolves at all in alkaline soil, or bonemeal, which is expensive and also dissolves slowly. Inefficient sources of potassium, such as organic matter or granite dust, are recommended in place of the much more efficient supplies of potassium used by the agribusiness.

An impartial observer comparing the two farming operations would begin to understand why synthetic fertilizers are preferred by most farmers in developed countries and will continue to be favored in the future, with no likelihood of a return to "the good old days" of sole reliance on organic matter additions. Yields on organic farms—such as those of the Old Order Amish, who eschew artificial fertilizers—are most often 10 to 40 percent less than on farms using conventional modern practices.

The world's burgeoning population will require even greater food production in the future, so yields will have to be as high as possible. Scientists have estimated that if most farms were converted to organic farming methods (no manufactured fertilizers or other chemicals used), from 20 to 40 percent more land would have to be cultivated to produce the same total crop yields now grown. More total soil erosion would occur because much of the remaining land available for farming is sloping land. Therefore, the universal adoption of organic farming techniques would be exceedingly hard on the environment.

What if the agribusiness managers use too much fertilizer? Excessive concentrations of chemical fertilizers, which generally have nutrients in the form of concentrated soluble salts, can damage or kill plants just as surely as if you poured a concentrated table-salt solution into the root system. Excessive nutrient levels may also interfere with normal plant growth: potato plants given too much nitrogen, for example, have luxuriant shoot growth but form very small tubers. In addition, the use of too much fertilizer may pollute ground water, rivers, and lakes.

To avoid these problems, the agribusiness managers can follow the recommendations of the county agricultural extension agent or have their soils and plants chemically analyzed for even more specific recommendations. In many states home gardeners, too, can send soil samples to state universities for testing that may be free or assessed a nominal fee. In states such as California where universities do not conduct this service, a private soil-testing laboratory should be contacted. Most scientists consider the home soil-testing kits offered for sale too inaccurate to be of much value. If you want to have your soil analyzed accurately, call your county agricultural extension director, who is listed in the phone book, for information.

Fairy Ring

5
A Mixed Bag of Water Wisdom

Soil with Hydrophobia

Those who try to maintain flawless lawns or golf greens are often bedeviled by "fairy rings"—circular strips of unsightly bare ground. Long ago these rings were a cause of real fear as well as annoyance. Some believed that they were created by diminutive supernatural fairies dancing in circles, which threatened great danger to the human passerby. The unlucky victim would be led inexorably into the ring by the bewitching fairy music and forced to join the fairies in their wild twirling dance. The dance might seem to its human victim to last only minutes, or an hour or two, but in fact its typical duration would be seven years or longer by our time.

Alternative explanations blamed gnomes and hobgoblins who buried their treasure within the confines of the ring, and dragons who breathed fire on the grass while at rest. More plausibly, it was suggested that the rings had been created by thunderbolts or whirlwinds, ants or moles, or that they marked the sites where old haystacks had stood.

In the last part of the eighteenth century the correct explanation was finally discovered. The rings were created by mushrooms, the best-known species being the "fairy ring" mushroom (*Marasmius oreades*), though a number of other mushroom species also form the rings. The familiar mushrooms we notice are only the "fruiting" bodies produced by the threadlike part of the fungus that grows underground. From the point of origin, the underground threads of the fungus spread outward at a rate of 3 to 13 inches in a year. At first, the fungus stimulates the growth of grass, which exhausts soil moisture; at the same time, the underground fungal bodies produce hydrophobic substances that prevent the rewetting of soil. (Though it is usually sandy soil that becomes hydrophobic, as sand particles are more easily coated with hydrophobic materials than the larger total of particle surface in clay and silt, the fungi can cause even soils with considerable clay and silt to become water repellent.) As the older part of the fungus in the center of the fairy ring dies out, the grass returns to that area while the circle of dead grass and mushrooms spreads outward. Some rings may attain a diameter of 300 to 800 feet and be from 250 to 600 years old.

People who have to deal with this problem in golf courses, lawns, or other areas where turf is important may increase water penetration by poking holes with mechanical devices, by replacing the hydrophobic soil and replanting, or by applying chemical wetting agents that allow the moisture to infiltrate readily. These agents, called surfactants, come in many forms with long chemical names—alkylpolyoxyethylene ethanol, polyoxyethylene ester of tall oil, etc.—and may be purchased at a plant supply company. They are expensive, however, and impractical for large areas, where they are difficult to apply properly. Fortunately, agricultural operations usually avoid the problems of hydrophobic soil because tillage mixes the less wettable with the more water-absorbent soil.

Water-repellent soil continues to float after wettable soil has sunk to the bottom of a glass of water.

Many other plant problems in sandy soils are due to their water repellency: patchy grass establishment on thousands of acres of Australian pasture, prevention of normal wetting under many citrus trees in Florida, and reduction of seed germination and establishment in areas where water cannot get to seeds or where greater runoff and erosion from water-repellent soil carries seeds away before they can sprout and grow. Gardens grown in sandy soils often have hard-to-wet areas.

Organic matter, although it can't be beaten as a soil conditioner and an aid to retaining water and nutrients, is strangely enough the principal cause of the hydrophobic soil. Some organic substances, leached from litter or created as microbial byproducts, can shed water when they are dry. If your garden soil becomes water repellent after applications of organic matter, you will have to take time to really soak it until the water

63

does penetrate. After that it won't be such a problem to wet the soil, especially if you don't let it become too dry.

Such factors as oil spills or even pets having a favorite sleeping spot will add an oil residue that repels water. Also, the heat from burning trash or leaves in the yard makes the repellency stronger.

Flooding Caused by Water Repellency

After brush or forest fires in hilly areas, downhill residents very frequently live in dread of rainstorms, which can bring disastrous flooding and mudslides into their yards and homes. Hydrophobic soil is considered to be one of the prime villains in this flow of debris. On the burned areas the surface is typically wettable, but in many places there is a layer of hydrophobic soil not far below the surface. Rainfall soaks in only a short distance before it is slowed or halted by the water-repellent layer; then the saturated upper layer is carried off as the water moves laterally along the repellent layer. The quantity of mud and other debris carried downhill can be enormous. For example, scientists measured 34 times more movement of soil and debris down a 50 percent slope following a moderately intense fire than on a similar unburned area. Rain-swollen streams carry the sediments to the more level valley floors below.

The layer of nonwettable soil under the surface layer in a burn area is formed by hydrophobic organic substances that are vaporized at the soil surface but condense on cooler, underlying soil. The best solution to the flooding and runoff problems would be controlled burning under optimum conditions—before excessive fuel buildup and when the soil is moist, so that temperatures and formation of water repellency are held to the minimum.[1]

1. See Leonard F. Debano, "Water Repellent Soils: A State-of-

A Mixed Bag of Water Wisdom

It is a lucky circumstance that water enters most soils with ease because a shortage of water is often the factor with the greatest potential to limit plant growth, especially now that the introduction of improved tillage, fertilizer usage, better crop varieties, insect and disease control, and other modern agricultural practices have reduced or eliminated many growth-limiting factors. Experiments have shown that high rates of nitrogen fertilization are ineffective when soil is allowed to dry to an extent that plants are water stressed. Phosphorus uptake may be limited by dry soil; therefore, less fertilizer phosphorus may be needed when the soil is kept at the proper moisture level.

When to Irrigate

With the increasing demand for water by expanding populations and economies, there is a growing urgency for efficient water use on agricultural lands where irrigation is needed. If the soil or the irrigation water is salty, some excess water above crop needs must be applied to leach the salts below the root zone, but generally it is best to apply only the amount of water that the crop actually needs. Excess irrigation water will not only be wasted but may also wash away nutrients, cause erosion of topsoil, flood the root zone and thereby cut off the supply of oxygen, or raise the water table (which may cause salt problems if dissolved salts move to the surface and become concentrated after evaporation).

You might think that the time to irrigate is when the plants begin to look dried out and droopy, and some plants show a color change when they start running

the-Art," United States Department of Agriculture, Forest Service, Pacific Southwest Forest and Range Experiment Station, General Technical Report PSW-46 (March, 1981).

low on water. But even well-watered plants may look droopy in the middle of a hot day, and when plants really do begin to need water, their growth usually slows down *before* there are any visual symptoms of water stress. Beans seem to be one crop that can be successfully watered by looking for a color change: if irrigation water is applied within a few days after the light green of the bean plants turns to a pronounced dark green yields will probably be only slightly reduced.

Another way to decide when to irrigate is by the feel of the soil. The soil in the root zone may still be wet when the surface is dry, so a shovel or a sampling tube is needed to go down at least six inches. A soil with a great amount of clay will easily form a ball in your hand when it has enough water, but when it takes some pressure to form a ball, it is time to irrigate. In contrast, a sandy loam soil that takes pressure to form a ball still has enough water; only when it won't form a ball at all does it need irrigation. Sand is the hardest to evaluate by feel, since it won't form a ball even when it does have enough moisture. If water is run at the proper time, when the soil is drying but not yet dry, water will reach depths of approximately one foot in clay soil, one and a half feet in loam, and two feet in sand per every inch of water applied.

The feel of the soil is useful, but it is still a rough estimate. A simple but accurate method of measurement to help farmers or gardeners determine the proper timing and amount of irrigation can be made with a relatively inexpensive instrument called a tensiometer. This instrument, which may be bought from specialty supply companies, is simply a tube that is inserted in the ground and kept filled with water; it has a porous cup at the end in the soil, and a pressure gauge at the aboveground end. The soil attracts the water from inside the tube through the porous cup; since the water-filled chamber is sealed, this will create a vacuum or suction (a pressure less than atmospheric pressure) that can be measured by the gauge.

The tensiometers should be installed at a minimum of two depths, one in the zone of maximum root density and the other near the bottom of the root zone. The bottom of the root zone may range from 18 to 24 inches for shallow-rooted crops such as cabbage, celery, corn, lettuce, potatoes, and spinach, and to more than 48 inches for deep-rooted plants such as asparagus, sweet potatoes, tomatoes, and watermelons.

TENSIOMETER

Research work has established the tensiometer readings (in the zone of maximum root activity) at which water should be applied for many common crops. A few days after watering, the tensiometer with its tip near the bottom of the root zone should show a change to a wetter reading, but not nearly as wet as the reading on the tensiometer in the zone of maximum root activity. This ensures that water has reached the bottom of the root zone, but not in sufficient quantities to cause large amounts of deep percolation. If water did not reach the lower instrument, the application time should be increased during the next irrigation; conversely, if excessive water is found in the low zone, the next irrigation should be decreased. The tensiometer system can also be used to trigger an automatic watering system.[2]

2. Despite the title, irrigation and water movement in soils are discussed in very easy to understand terms in S. A. Taylor, *Physical Edaphology: The Physics of Irrigated and Non-Irrigated Soils* (San Francisco: Freeman, 1972).

Common Misconceptions about Water and Soil

One obvious cause of water loss is evaporation, and in soil made up of a large concentration of clay particles, water can readily move as far as a couple of feet up to the evaporating surface. Water films crawl up through the fine intervening soil pores like kerosene moving up the wick in a lamp. In coarse sandy soils, by contrast, the relatively wide distances between particles reduce the wicklike action and consequent evaporation. This can have some most surprising results.

Desert sands allow the scanty rains to evaporate back into the atmosphere within a short time and are again arid—right? Wrong! If you dig down a foot or so in desert sand dunes, you can almost always find sand wet enough to make sand castles. The decreased capacity for evaporation in sand will also cause a pile of sand in your yard to dry out much more slowly than a pile of clay. So even though sand has less capacity to hold water in quantity than does soil with finer particles, the smaller ability of sand to act like a wick prevents much of the water from rising to the surface and evaporating.

There are misconceptions about water moving downward as well as upward. Suppose you have just purchased an expensive shrub and have read the instructions carefully on the care and planting of this species. They tell you that good drainage is especially important. You dig a large hole (the type called a $100 hole for a $20 shrub), and to be on the safe side you spread a layer of coarse gravel at the bottom before putting in the plant and soil. In this way, you feel sure, the water will drain well and the roots won't stay too wet.

Have you done the right thing? Actually, you may have harmed your new plant. Since impervious layers such as hardpan greatly slow down water movement, it seems logical to assume that underlying, open, coarse soil will improve drainage. But this is wrong because gravelly or sandy layers act to slow down water movement if the overlying soil has finer particles. The finer soil above, with its greater particle surface area and smaller pores, has a greater attraction for water than the coarse particles, and water cannot move *down* into the coarse soil by wicklike action. Water must build up until it is so wet that gravity will finally begin to pull some moisture into the coarse layer.

An analogy can help to clarify what is happening. If you pour a cup of water into gravel, it will quickly run through it. If you pour the same amount of water into sponge (representing the greater water-holding capacity of finer soil) sitting on the surface of the gravel layer, it will soak the sponge *without entering the gravel below*. Water would have to be supplied in amounts exceeding the water-holding capacity of the sponge before it would enter the gravel through the force of gravity. But if the same amount of water is poured on the same-sized sponge resting on several other sponges (representing a soil with the same fine particles and high water-holding capacity throughout its depth, with no subsurface layer of coarse material), the water will be absorbed on down through the greater depth of sponges, and nowhere would any sponge be as wet as the sponge resting on gravel. In one case a horse-racing track was built with soil spread over coarse material to improve drainage. This caused the track to remain so wet that the coarse particles had to be dug up and replaced with other soil.[3]

In another way, gardeners have at times unwittingly gone awry in their attempts to use water wisely when

3. Personal communication, Dr. Lewis Stolzy, University of California at Riverside.

"gray water" from tubs, washers, and sinks, has been used for plant watering. This water, like water treated by zeolite water softeners, has a high, and in many cases, harmful sodium content. The sodium can cause burn in sensitive plants such as orchids, avocado, or citrus. Even more important, it can cause soils with large amounts of clay to have much lower rates of water penetration. The sodium makes the soil particles separate from one another and clogs up the larger water-draining pores.

Poorly draining soils with large amounts of sodium are called "sodic" or "black alkali" soils. Their dark color comes from humus particles that are separated and dispersed by sodium in the same way as the clay. Black alkali soils are found in some scattered and infertile locations of the West and Midwest, where the sodium comes from irrigation water or is already present in the soil. The soils are often located in small areas called "slick spots," surrounded by more productive soils. Only the most tolerant plants, if any, can grow on the black alkali soils. The problem can be overcome by replacing the sodium with calcium (generally added in gypsum form) and then leaching out the displaced sodium. The calcium allows soil particles to have greater attraction for one another and aids in the formation of soil granules that permit easy penetration of water through their intervening spaces.

Even more serious to plants can be the boron from borax contained in some soap and detergents. Plants need a small amount of boron to grow, but even a small excess can cause their injury or death.

Agricultural Drainage

Establishing good drainage of agricultural lands can sometimes be a very expensive project. The rich farms in the Imperial and Coachella valleys of the Southern

California desert are underlain by huge numbers of drainage tile lines, placed there to carry away excess salt-rich irrigation water. Subsurface impervious layers in the valley soils slow water movement to such an extent that saturated layers—called "perched water tables"—build up below the surface. Then, when the irrigation water moves to the surface and evaporates, it leaves the salt behind. The buildup of salt would soon kill even the most salt-tolerant crops were it not for the tile drains, which allow the salt to leach through the soil to the drains and flow into the nearby Salton Sea.

On some other California farms, however, bad drainage is eagerly sought. In the ricefields of the Sacramento and San Joaquin valleys, which have dense clay soils or impervious hardpans that prevent the downward percolation of water, remarkably high yields of rice are achieved by growing improved varieties, spreading seed and herbicide from airplanes, and fertilizing with plenty of nitrogen. A high-technology method is used to level the fields: a rotating laser beam mounted on a mast establishes a reference elevation that can be detected by earthmovers over a substantial distance. The reference beam allows such precise leveling that after levees are constructed, water will stand no less than two inches deep adjacent to the uphill levee and no more than four inches deep adjacent to the downhill levee. The precise water depth, maintained throughout the growing season by inlet and outlet boxes in the levees, works in combination with herbicides to decrease infestations of noxious weeds.[4]

Deadly Percolation

When soil water, instead of moving down through the soil, flows strongly upward, a frightening phenomenon

4. See J. Neil Rutger and D. Marilin Brandon, "California Rice Culture," *Scientific American* 244, no. 2 (February, 1981): 42–51.

can develop—quicksand. The secret of quicksand is not a mystical property of the type of sand but an upward flow of ground water holding particles in suspension, so that even though the soil looks like solid ground, it is in effect a heavy liquid. Thus the suction or pull of quicksand is the force of gravity in a heavy liquid with viscous properties, which create drag on a terrified body. If you fall into quicksand, you will float better than in water; if you have this knowledge and confidence, you can slowly swim to solid ground.

Quicksands often occur in river or lake beds at points where ground water emerges, flowing upward under artesian pressure. Quicksands sometimes develop also on the downstream side of an earth dam after the reservoir is filled and high seepage pressures build up in the soil beneath the dam. This occurs frequently on the land side of levees along the lower Mississippi River during flood stage.

Because quicksand is caused by seepage forces, the soil can be made capable of providing support by eliminating the seepage pressure. Quicksands may exist in some areas during a time of the year when there is substantial artesian water flow and not at other times, when the flow is reduced. Many quicksands are made up of fine sand particles, but sometimes the condition may not even involve sand. "Quicksands" may be composed of the finer silt particles; if the flow of water is strong enough, they may even contain suspended gravel fragments.[5]

5. Quicksand is discussed in G. S. Spander and R. L. Handy, *Soil Engineering* (New York: Harper & Row, 1973), pp. 273–75.

Dust Cover

74

6
Rock on Its Way
to the Ocean

What does erosion have in common with the fall of a
ripened apple to the ground; the rising and falling tides
at the shore; the rounded shapes of stars, planets, and
moons; and the orbit of the moon around the earth?
They are all dependent on gravity. Because of gravita-
tional attraction, soil and rock fall and slide down steep
slopes; glaciers flow outward and shear off land surfaces; raindrops pulverize dirt clods and then carry
detached soil particles in the runoff water that gravity
pulls downhill. Gravity is even the driving force of wind
erosion, because wind results from the flow of denser
air masses beneath masses of thinner air.[1]

Landslides

Landslides, the most direct effect of gravity on soil
movement, provide some of the most calamitous examples of erosion. In Kansu, China, near the Tibetan bor-
der, a cliff 100 feet high composed of loess once housed
thousands of peasants in carved-out cave dwellings.
The loess bands of the cliff had formed over eons from

1. An interesting discussion of erosion is found in J. Gilluly,
A. C. Waters, and A. O. Woodford, *Principles of Geology*, 3rd ed.
(San Francisco: Freeman, 1968), pp. 60–87.

the yellowish silty dust carried on the wind from the Gobi Desert. For one of those peasants, homeward bound with his load of firewood for cooking and warmth, there would be no thought of disaster—only the anticipation of the nighttime meal and rest in the hospitable cliff that sheltered his family and the great number of others who found homes there. He could not know that on December 16, 1920, an earthquake would topple the face of the cliff into the Hwang Ho river valley, burying him and his family. More than 180,000 people died, including most of the cliff dwellers and victims in the buried cities and villages of the river valley. That day of terror is known in China as *Shan Tso-liao,* or "when the mountains walked."

The violent shaking of a powerful earthquake can transform water-saturated silts and sands from a solid to a liquidlike state. This process, called liquefaction, can cause the granular materials to flow like a liquid on even gentle slopes. The pulsating, liquefied sediments may also turn into quicksand, on which buildings may tilt and sink, while buried tanks may rise buoyantly.

Earthquakes can play some other startling tricks on the soil. Earth waves, very similar to ocean waves, may race across the ground, usually in places with a large amount of loose alluvium or fill. Waves of six inches to a foot are not uncommon during large earthquakes, but reports of four- to six-foot crests probably more often result from the overwhelmed imaginations of badly shaken observers.

Another curious effect is the creation of small craters by upward explosions of sand and water during violent earthquakes. After the 1906 San Francisco earthquake, such craters were commonly found in the flat alluvial plains near Watsonville, California.

Landslides may be triggered not only by earthquakes but by cuts—such as those made for highways or railroads—that reduce support for the slope above, or even by removal of timber. Along the Pacific coast, from northern California to Alaska, some of the best land for productive forests occurs on unstable slopes. Strong roots of trees and shrubs on these slopes help prevent the soil from sliding down the hillsides; vertical roots anchor the soil mass to cracks in the bedrock, and horizontal roots interlock to tie the hillside together laterally. But after forests are cleared, the dead roots begin to decompose. Several years later, when the rotting root material has lost much of its reinforcing strength, the soil mantle can slide down the hill. The landslides may be prevented if enough vegetation is left to anchor the soil.[2]

Although highways or railroads can generally be rerouted to bypass a landslide area, there have been cases where structures threatened by sliding earth could not be moved, and very ingenious methods were found to stop or slow down the movement. An oil field near Ventura, California, for example, was on a slide area that slowly bent or even sheared off oil well casings as soil moved downslope. Movement was most rapid in the winter when rains saturated the ground, changing the clays into a lubricant. To avoid abandoning the valuable oil field, the entire hillside was paved with asphalt, passageways were bored or dug along the base of the slide, and tile was installed to carry off seepage from the rains. Thereafter, most of the rain ran off the asphalt pavement, and the small amount of moisture that did penetrate was drained away. The slide was stopped by stopping the lubrication of the clays.

When the enormous Grand Coulee dam was being built on the Columbia River, a tremendous amount of

2. The relationship between roots and slope stability is evaluated by R. R. Ziemer in the *McGraw-Hill Yearbook of Science and Technology* (New York: McGraw-Hill, 1981), pp. 342–44.

saturated silt began to creep into the excavation at the dam site. To halt the impending slide, the engineers developed an extremely clever plan: the silt was perforated with many pipes in which a refrigerant was circulated, thus freezing the water in the pores of the silt and cementing its particles. The air was kept refrigerated, and the increased strength of the slope stopped the movement until the concrete for the dam was poured. Once the concrete had hardened, the restored support at the base of the slope stabilized the soil above.

Is Erosion of U.S. Farms Leading to Calamity?

The erosion of most concern is understandably that on farmlands. The best estimate of the U.S. Department of Agriculture is that the most severe soil erosion is concentrated on a relatively small proportion of U.S. farmland: 70 percent of all excess soil loss, for example, is estimated to come from only 8.6 percent of the cropland.

Unfortunately, this land includes some of the best-producing areas, in states such as Iowa, Missouri, and Mississippi, as well as Kentucky, Tennessee, and the Southeast. The USDA estimated that the potential loss of crop yield over the next 50 years, assuming current rates of erosion, would be equivalent to the output of 23 million acres. Out of an estimated 540 million acres of currently used or potential cropland in the country, this loss may not be disastrous, but it is certainly not insignificant.[3]

How were these estimates made by the USDA, and

3. Comprehensive estimates of soil erosion in the United States are found in USDA, *Basic Statistics, 1977 National Resources Inventory* (Washington, D.C.: Soil Conservation Service, 1980). See also Pierre R. Crosson and Anthony T. Stout, *Productivity Effects of Cropland Erosion in the United States* (Baltimore, Md.: Johns Hopkins University Press, 1983).

how accurate are they? Water erosion and wind erosion losses are not actually measured but calculated using equations derived from enormous amounts of data collected by scientists to relate measured erosion from fields to farm and climatic properties.

There are inevitably some uncertainties in the erosion estimates:

1. The equations were designed to predict erosion for well-characterized areas, and using them to calculate erosion over enormous expanses of U.S. farmland leaves room for error.

2. The rate of soil formation is estimated by measuring the depth of soil that has accumulated over material of known age (often determined by radiocarbon dating), but such measurements have been made in relatively few locations, and the rate at which the topsoil formed may not have been constant (if climate change occurred, for instance).

3. There is also some uncertainty about the effect of erosion on crop production.

Even so, there appears to be plenty of cause for concern. Taking all measures that will help prevent erosion is logical and necessary.

There are many good techniques to reduce soil erosion: "conservation" tillage — minimum-till or no-till — diminishes the possibility that soil will be loosened and easily erodible (nontillage is discussed in chapter 7); steep lands can be terraced; crop residues can be left in place; contour plowing can be used; and runoff can be collected in basins. These techniques will also reduce water pollution from eroded sediments (which fill up reservoirs, lakes, and channels) and from fertilizers, herbicides, and pesticides.

Farmers often avoid using erosion control methods because they believe that these strategies don't pay for themselves. Yet methods of conservation tillage that reduce soil loss may result in lower labor and fuel costs than conventional tillage. Since the 1960s various forms of conservation tillage have spread widely; they are now practiced on an estimated 25 percent of cropland, and that percentage may well increase significantly in the future. But problems of erosion will continue; large areas not now used for crop production may be brought under cultivation as the need arises, and much of the additional land would be more erosive than most territory now in production. Potential farmland that is now pasture or forest will be so prone to soil loss when converted to cropland that vigorous attempts at erosion control will be essential for its successful use.

Why does substantial loss of topsoil have such a bad effect on crop growth? Topsoil has higher amounts of organic matter and nitrogen and also sometimes more phosphorus and potassium than subsoil; therefore, it has greater fertility. Subsoil can often be converted to topsoil by adding chemical fertilizers and large amounts of organic matter, but changing subsoil to topsoil is much more practical in gardens than on large farms.

The productive capacity of U.S. cropland is probably being lost at an even faster rate by conversion of farmland to other uses than by the forces of erosion. According to the best available estimate, made by Michael Brewer and Robert Boxley for the federally sponsored National Agricultural Lands Study, about 875,000 acres of actual or potential croplands were converted to urban uses each year between 1967 and 1975.[4] No one knows what the future rate of conversion will be, but if

4. Michael Brewer and Robert Boxley, "The Potential Supply of Cropland," in *The Cropland Crisis: Myth or Reality?* ed. Pierre Crosson (Baltimore, Md.: Johns Hopkins University Press, 1982), pp. 93–116. See also discussion of conversion to non-agricultural

the rate of farmland loss to urban uses stays the same, 21.9 million acres of actual or potential cropland will be removed between 1975 and the year 2000.

Dust Bowl Disasters

Although water erosion removes more soil from farmland than wind erosion, it was wind erosion in the dust bowl of the southwestern Great Plains that resulted in the establishment of the Soil Conservation Service, with large commitments of money and federal personnel, to protect the soil. Hugh Bennett, who would soon become the first director of the Soil Conservation Service, was testifying before a congressional committee in 1935 on the need for a large program to control soil erosion. During his testimony a dust storm filled the air over the nation's capital with brown gloom, dimming the sunlight. "This, gentlemen," he announced, "is what I have been talking about."[5] Congress quickly passed legislation establishing the Soil Conservation Service, with Bennett at the helm.

When the Great Plains were settled, the natural grass cover was plowed under, and the soils were planted with corn and wheat. In the early 1930s a succession of dry seasons led to the appalling drought years of 1933 and 1934. The pulverized, tilled soil had not only lost its protective grass cover but much of its cohesiveness, and in the following years the winds brought the choking dusters. In May of 1934 powerful blasts of wind lifted dust estimated to total more than 300 million tons, enough to fill six million railroad cars, from the

uses, in Pierre R. Crosson, "Future Economic and Environmental Costs of Agricultural Land," ed. Pierre R. Crosson (Baltimore, Md.: Johns Hopkins University Press, 1982), p. 179.

5. Donald Worster, *Dust Bowl: The Southern Plains in the 1930's* (Oxford: Oxford University Press, 1979), p. 213.

Dust cloud rolling over a town in western Kansas, February, 1935.
Photograph courtesy Kansas State Historical Society, Topeka.

fields on the Great Plains. This "black roller" churned
eastward, gagging people and animals and changing
day into darkness. As the black blizzard moved across
the well-watered lands east of the Mississippi, it picked
up very little loose soil from the forested and grass-
covered landscape, and the dust started falling to the
ground—more than 100 tons per square mile. The dirty
clouds still contained enough dust, though, to dim the
sun as they reached New York City, and the prairie soil
kept blowing eastward over the Atlantic, dropping dust
on decks of ships as much as 300 miles offshore.

The dust storms continued for years, through the
"dirty thirties," not decreasing in frequency until heav-
ier rains finally began to overcome the drought in the
late 1930s and early 1940s. The worst year on the plains
was probably 1935, when dust storms blew almost
every day for six weeks from early March to mid-April.
A Kansas woman described her experience as follows:

All we could do about it was just sit in our dusty chairs, gaze at each other through the fog [dust] that filled the room and watch the fog settle slowly and silently, covering everything—including ourselves—in a thick brownish gray blanket. When we opened the door swirling whirl winds of soil beat against us unmercifully—. The door and windows were all shut tightly, yet those tiny particles seemed to seep through the very walls. It got into cupboards and clothes closets; our faces were as dirty as if we had rolled in the dirt; out hair was gray and stiff, and we ground dirt between our teeth.[6]

As the dust storms followed one after another, some people wondered if they would ever end. Woody Guthrie sang:

It fell across our city like a curtain of black rolled down.
We thought it was our judgment, we thought it was our doom.[7]

Some plains inhabitants took the disaster less seriously. Children often enjoyed the chaos created by the storms, particularly when schools were closed, and newspaper editor William Allen White called the dusters "the greatest show" since Pompeii was buried in ashes.[8] The combination of dust bowl and Depression was too much for other people, though, and many—such as the "Okies" depicted in John Steinbeck's *Grapes of Wrath*— moved on to new locations. In the words of Woody:

6. Ibid., p. 17.
7. Woody Guthrie, "The Great Dust Storm." Words and music by Woody Guthrie, TRO copyright 1960 and 1963, Ludlow Music, Inc., New York, N.Y. Used by permission.
8. Worster, *Dust Bowl*, p. 17.

We loaded our jalopies and piled our families in,
We rattled down the highway to never come back
    again.[9]

There were ten thousand abandoned houses on the high plains and nine million abandoned acres of farmland.

Most of the soil particles moved by wind in the dust bowl did not travel across the country. Only the finest particles in a dust storm go far. The impact of the sand grains that roll or skip along the surface causes other sand grains to be shot off the ground, and all these bouncing grains sandblast the finer dust particles into the air to create the seething dust clouds. The sand travels anywhere from a few yards to a few miles, forming dunes, filling up hollows, and drifting against farm buildings and hedges. For every ton of the far-ranging airborne dust, generally two or three tons of sand will pile up in dunes or drifts close to the source. In the years since the mid-1930s, rainstorms have almost completely obliterated the eroded hollows and drifts of sand and silt made by those dirty-thirties dust storms. Even in areas that are very susceptible to wind erosion, water often removes greater amounts of soil than the wind.

The dust bowl disasters led to increased use of methods for control of wind erosion: planting trees for windbreaks; alternating strips of fallowed land with protective strips of crops; leaving the previous season's stubble and crop remains on the soil surface for protective cover; and tilling the soil so that the surface is roughened (which reduces wind velocity and traps some of the moving particles). If a drought lasts for several years in a nonirrigated area, however, protection becomes extremely difficult or impossible, because

9. See note 7.

all protective cover will disappear. Therefore, when drought returned to the southern plains in the "filthy fifties," the dust storms returned as well, and another drought in the mid-1970s brought still more dusters.

Deep-well irrigation has now removed the danger of drought and dust storms for many farmers on the plains. Much of the recent increase in irrigated agriculture has been in dust bowl states, particularly Nebraska, Kansas, Oklahoma, and Texas. Here irrigation is used mainly to grow corn, sorghum, and alfalfa, mostly for nearby cattle feedlots. The irrigation water comes from the Ogallala aquifer, a gigantic underground lake that spans eight states. Eventually, a falling water table and more expensive energy for pumping will probably make irrigation less popular, and an increased danger of dust storms may reappear.

Great man-made dust storms have not been limited to the American continent. In 1954, Soviet premier Nikita Khrushchev initiated his "virgin lands" scheme to cultivate 100 million acres of previously untilled land in Soviet Central Asia and Siberia. In June, 1960, after severe drought, the area suffered gigantic dust storms.

Good Effects of Erosion

It is easy to write bad things about erosion, but we should recall the very apt saying that "it is an ill wind that blows no good," for the eroded material can have good effects in areas where it is deposited. For instance the "black earths" of the Great Plains, discussed in chapter 2, the richest of all U.S. soils, have developed from windblown loess deposits; and sediments left by streams or rivers have created many fertile soils, especially in such floodplains as those along the Mississippi River. In addition, the eroded soils deposited by rivers at the margins of continents have created huge areas of valuable real estate, widening the continents by as much as 500 miles.[10]

As the continental glaciers advanced during the great ice ages, trapping enormous amounts of water in their frozen ice sheets, the dropping sea level exposed vast additional areas of normally offshore sediments. Fifteen thousand years ago, during the last ice age, this process caused the sea level to drop by about 425 feet. Areas that are now sea floor were covered by forests, meadows, ponds, and marshes; birds once flew among the trees in many places where fish now swim.

Although new land may be created at the coast by sediment deposits or soil upthrusts, land is also lost by the erosive action of the waves. In the early 1900s, outraged property owners in England demanded that the government take action to stop the loss of their land to wave action. A Royal Commission, appointed to study the matter, concluded that over a period of 35 years England and Wales lost 4,692 acres from wave erosion but gained 35,444 acres, for a net gain of almost 900 acres a year; clearly, people whose land had disappeared complained a lot more loudly than those whose land was expanding. Beachfront property owners should realize that the very existence of a cliff indicates ongoing erosional processes; they must exercise caution in choosing where to build, lest their homes be lost to the waves.[11]

Sediments Claimed by the Ocean

Some southern California beaches have shrunk badly because eroded sediments have been prevented from reaching the coast. South and west of the Los Angeles

10. K. O. Emery, "The Continental Shelves," in *The Oceans*, ed. Dennis Flanagan (San Francisco: Freeman, 1969), pp. 39–52.

11. See Willard Bascom, *Waves and Beaches: The Dynamics of the Ocean Surface* (New York: Doubleday, 1964).

plain, the beaches were in the past supplied with sand carried from inland areas by small, intermittently flowing streams. But with the demand for flood control, the streams have been dammed and their channels lined with concrete. Now the sand is trapped behind the inland reservoirs. The undersupplied beaches retreat as the beach sand is swept southward along the shore zone by wind-driven waves and their associated currents, which finally dump it into the Newport submarine canyon from which it cannot be recovered. Instead, great quantities of sand are periodically dredged from the sea floor and laid on the beaches to widen them.

Erosion does not stop at the edge of the ocean. Submarine canyons, some deeper and more precipitous than the Grand Canyon of the Colorado River, are found at depths far below the maximum probable lowering of the sea level during the ice ages. This means that the canyons were formed by some process of erosion that operates under water. Apparently, when sediment slips downslope, it becomes mixed with the overlying water, increasing the density of the water and giving it an impetus. This causes the water to flow rapidly downslope and erode the weakly consolidated sediments over which it glides.

Eroded soil materials have not only accumulated on continental margins but spread throughout the oceans. Most of the very deep ocean floors are covered with red clay. The red color is from the rusting of iron contained in the soil. Because the minerals of the red clay are like those formed on land, it is thought that most of this clay has come from the continents, transported by winds, rivers, and ocean currents. Dissolved chemical substances, released when terrestrial rocks were weathered to form soils and carried to the ocean by rivers, have provided most of the salt in seawater and a large part of the essential nutrients needed by sea life.

Plowing

7
The Plowman's Folly

Evolution of the Plow

One of the oddest features of the European landscape is the contrast in the shape of fields in different geographic regions. Cropland is arranged in long strips around villages north of the Loire River in France, in the Netherlands, Belgium, West Germany, and—in previous times—in England and southern Scandinavia. Conversely, fields are more nearly square around the smaller hamlets of southern France, western Britain, and northern Scandinavia. The cause of these differences in field shapes is still not known for certain, but according to a commonly held theory, they resulted from the use of different types of plows.[1]

The square plots probably originated with the ancient "scratch plow," which has been used for tillage since Sumerian times around 5,000 years ago. The primitive form of the scratch plow is a heavy pointed stick (in more advanced forms it has a metal point), which is usually pulled by two yoked oxen. The scratch

1. The theory about the relationship between field shapes and the use of different types of plows is discussed in D. G. Grigg, *The Agricultural Systems of the World: An Evolutionary Approach* (Cambridge: Cambridge University Press, 1974), p. 160.

SCRATCH PLOW WITH OXEN

plow has remained the dominant type in southern Europe, North Africa, India, and most of Asia. It normally does not turn the soil over but parts the earth, throwing up loose dirt on each side of a shallow furrow and leaving a wedge of undisturbed soil between the rows. Therefore, after the field has been plowed in one direction it is necessary to plow at right angles across these furrows to ensure that all the soil has been tilled — and thus square fields are most suitable.

Although the scratch plow worked well wherever the soil was easily cultivated, it was not adequate for the firmer, hard-to-work earth that covers much of northern Europe. The soils there tend to have larger amounts of clay than those of the drier regions to the south, because the wet climate of northern Europe causes greater rates of rock mineral weathering and clay formation. By the sixth century a new heavyweight plow was introduced to northern Europe, probably by barbarian invaders. Known as the "moldboard plow" in its fully developed form, the new tool could cut more deeply into the clay-rich soil and turn the cut slices of turf or topsoil over to one side.

The heavy plow was much more difficult to pull than the scratch plow, requiring four or more oxen rather than a team of two. As a result, it was more practical to plow a longer strip before turning the team, and it may be that the long, unfenced fields in northern Europe originated with the use of the heavy moldboard plow.

The moldboard made possible the cultivation of great areas of former virgin forest land. Food production increased and so did population as use of the new plow spread gradually over northern and western Europe. However, the expense of making and maintaining the new type of plow, and the large team of oxen needed to pull it, meant that only rich men could own their own. Peasant farmers had to band together in joint ownership of a plow and team, and they had to merge their plots of land into the long strips of open, jointly worked fields. The medieval manor evolved around this basic framework.

Why weren't horses used to pull these early versions of the moldboard plows? Horses' hooves fared badly on the wet soils of northern Europe, and the horse did not appear to be as strong as the ox. Around the eighth or ninth century the invention of the horseshoe to protect the hooves and the use of the horsecollar harness greatly increased the horse's pulling power. Previously, the animal had been half choked by the harness strap over its windpipe every time it strained forward to pull a load. With the introduction of horseshoes and the horsecollar, horses proved greatly superior to oxen for pulling heavy plows; they were found to have as much strength, more endurance, and much greater speed than oxen, although they were more expensive to feed. The horse replaced the ox slowly in northern Europe,

HORSE-DRAWN MOLDBOARD PLOW

where oxen still outnumbered horses in the eighteenth century, and made even less progress in southern Europe, where less pulling power was required for the scratch plows still used in the sandier soils.

In North America the plow remained less important for cultivation than the hoe until the nineteenth century, and in the southern United States the hoe was the dominant tool used by slaves or poor farmers until the middle of the nineteenth century.

Thomas Jefferson, a farmer as well as third president of the United States, was the first American to make a careful study of the plow.[2] He found that it was very important to make the cutting edge a straight line, but he did not seek a patent. The first patent on a plow was given in 1793 to Charles Newbold of New Jersey, who developed a cast-iron model. Newbold lost all his money rather than making a fortune, however, because neighboring farmers refused to buy his plows, many of them thinking that the iron would poison the ground.

In 1819, Jethro Wood obtained a patent for an iron plow with interchangeable parts. This design, the nineteenth to be patented in America and based on Jefferson's theories, was much more successful than Newbold's. Wood sold every plow he could make, even though he had a lot of trouble with patent infringement.

Jethro Wood's plow worked well in the relatively sandy soils of the eastern states, but it ran into difficulties in many soils farther west—for example in Illinois, Indiana, and Wisconsin. Farmers grumbled that plowing these soils was like trying to plow a mixture of mud, tar, and molasses. The farms in the new lands were large; to allow a farmer to plow a fast, clean furrow the

2. The improvement of plows is traced in Percy W. Blandford, *Old Farm Tools and Machinery: An Illustrated History* (Fort Lauderdale, Fla.: Gale Research, 1976), pp. 43–73.

muds had to fall away from the moldboard in a smooth curl, a process called "scouring." Sticky clay soils in Illinois did not scour but stuck in thick gobs to the iron, forcing a plowman to stop repeatedly and clean off the plow's surface. Men of inventive mind thought these problems could be solved by making changes in the shape of one or several parts of the plow, and patents were granted for various designs.

Enter the hero: John Deere, a young blacksmith from Vermont who moved to Illinois in 1836, saw the problems of plowmen and reached the conclusion that the difficulties resulted from the iron itself rather than its shape. While visiting a sawmill, Deere saw how a steel saw blade had been polished shiny by the friction of cutting, and wondered whether steel would also clean itself when cutting through earth. He made a plow out of a discarded circular steel saw, pounding it to shape with a wooden mallet so it wouldn't be dented, and tried it out in the field of a farmer named Lewis Crandall. Deere cut a dozen smooth, straight furrows without having to stop once to clean the plow. His neighbors continued plowing on down the field just to convince themselves that this incredible improvement was real.

Deere sold his first plows for ten dollars each. As the blades slid smoothly through the ground, they vibrated with a humming sound, and John Deere became famous for his "singing plow." When settlers pushed farther west to new lands such as those in California or Oregon, John Deere plows were among the most valued cargo in the wagon trains.

John Deere's business grew large and prosperous, and by 1869 his factory office in Moline, Illinois, was a showplace of the Midwest, with gas chandeliers, central heating, silver-plated locks and doorknobs, and a frescoed ceiling. It must have been a Deere salesman that Mark Twain quoted in *Life on the Mississippi*: "You

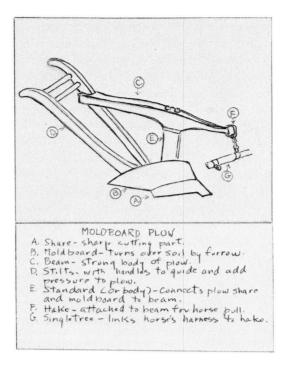

MOLDBOARD PLOW
A. Share- sharp cutting part.
B. Moldboard- turns over soil by furrow.
C. Beam- strong body of plow.
D. Stilts- with handles to guide and add pressure to plow.
E. Standard (or body)-connects plow share and moldboard to beam.
F. Hake- attached to beam for horse pull.
G. Singletree - links horse's harness to hake.

show me any country under the sun where they really know how to plow, and if I don't show you our mark on the plow they use, I'll eat that plow, and I won't ask for any woostershyre sauce to flavor it up with, either."[3] Today, of course, the John Deere farm machinery company, with its large line of tractors and agricultural implements from small to gigantic, is America's largest farm equipment manufacturer, with sales totaling billions of dollars a year.

A farmer in the 1840s with a steel plow had done a good day's work if he finished plowing an acre by sundown. An acre a day remained the quota for most

3. Mark Twain, *Life on the Mississippi* (New York: New American Library, 1962), pp. 328–29.

farmers in the United States throughout the rest of the nineteenth century and the early part of the twentieth. Some farmers in the later part of the nineteenth century used steam tractors, which could cover ten times the area that horses could plow in a day, but these puffing monsters were so cumbersome and expensive that their use was not widespread. Gasoline-powered tractors began to replace horses in the period between World War I and World War II in the United States, and after World War II on the generally smaller farms of Europe.

It may seem that a very romantic part of farming was lost when most farmers stopped relying on their living form of "horse" power.

> To plow and to sow,
> And to reap and to mow,
> And to be a farmer's boy.[4]

Bucolic verse, such as this sample from an anonymous poet, fits the vision of freshly turned earth and men with their horses in simpler times better than the efficiency of the tractor age. But after farmers learned that they could plow 10 acres a day with their new machines compared with the acre a day they could till with their teams of horses, few wanted to return to the old ways. Now, farmers with advanced systems can plow 50 acres or more in a day as they sit comfortably listening to radios in the air-conditioned cabs of their $65,000 tractors.

Although the moldboard plow has played a role of great importance in agricultural and economic expansion, it has also received some severe criticism. Why would the plow—almost as hallowed a symbol of American history as motherhood, apple pie, and the Fourth

---

4. Anonymous, "A Farmer's Boy," quoted in Bergen Evans, ed., *Dictionary of Quotations* (New York: Bonanza Books, 1968), p. 226.

of July—ever be considered the cause of great harm? The plow has been blamed for the loss of vast amounts of soil because plowing creates fields of bare, loosened earth that are vulnerable to water and wind erosion.

In 1943, Edward H. Faulkner, a farmer and a county agricultural agent in Ohio, devised one of the first farming systems that avoided plowing. He substituted disks for plows (disks do not go as deep and turn the soil less), and left some crop residue on the surface for protection from erosion. Faulkner discussed his ideas in *The Plowman's Folly,* published in 1943; the book got heated reactions from agriculturalists who regarded plowing as synonymous with farming.[5]

In addition to Faulkner's system, other reduced tillage or no-till methods have been proposed—some serious, some fanciful. A New Yorker, pointing to a hillside field, complimented the New Englander on his corn.

"How do you plow that field? It looks pretty steep."

"Don't plow it; when the spring thaws come, the rocks rolling down hill tear it up."

"That so? How do you plant it?"

"Don't plant it really. Just stand in my back door and shoot the seed in with a shotgun."

"Is that the truth?" asked the New Yorker.

"Hell, no. That's conversation."[6]

Farming without Plowing

On a large and increasing amount of farmland in recent years, crops are being grown without any tillage

5. The title of this chapter is taken from Edward H. Faulkner, *The Plowman's Folly* (New York: Grosset & Dunlap, 1943).

6. Anonymous joke in *Fun Fare: A Treasury of Reader's Digest Wit and Humor* (New York: Simon & Schuster, 1949), p. 215.

whatsoever.[7] A narrow slit is cut in the soil by a planting machine that also drops in seeds and chemical fertilizers, and a wheel at the back end of the machine then presses the soil down over the seed. Therefore, most of the soil remains covered with the residues of old crops. Weed control, a major role of plowing or disking, is achieved partly by herbicides and partly by the mulching effect of the old crop remains.

If non-tillage methods are used, erosion can be reduced a hundredfold or more because of the protective mulch cover; the amount of fuel needed by farm equipment to establish a crop can be cut by as much as two-thirds compared with conventional tillage; and only a third as many working hours are required. In addition, non-tilled soils are frequently more productive than conventionally tilled acreage because the organic surface mulch conserves moisture, and the absence of tillage prevents excessive breakdown of the aggregated clumps of soil that allow easy penetration of water, air, and roots. Corn yields from non-tilled, well-drained soils are generally 10 to 20 percent higher than those from tilled soil.

According to one projection, by the year 2010, 50 to 60 percent of the United States' cropland will be farmed by methods that use reduced tillage or non-tillage and that cause less disturbance of plant residues than conventional tillage.[8]

Plows and Soil Compaction

You may be surprised to learn that plowing the soil can cause subsurface layers to become compacted and re-

7. Ronald E. Phillips and Shirley H. Phillips, eds., *Non-Tillage Agriculture: Principles and Practices* (New York: Van Nostrand Reinhold, 1984).
8. Pierre R. Crosson, "Future Economic and Environmental

sistant to passage of roots, water, and air. At the deepest level of cultivation, hard and nearly impervious layers may be formed where repeated use of a plow compresses pulverized soil particles against the underlying uncultivated soil. This "plow pan" makes a strong barrier against root growth and moisture movement, and can prevent plants from getting water and nutrients from lower soil depths. If a strongly compacted plow pan forms at a depth of one foot, for instance, plants are limited to one foot of growing space, where they might otherwise have roots growing down several feet.[9]

Impervious plow pans can be created without heavy equipment. On fields in Poland and India that have never been mechanically farmed, plowing compaction problems have been observed that are as bad as cases on land farmed with heavy machinery in the United States. Even walking can severely compact soils, creating problems in such locations as golf and bowling greens. These sites of compacted turf can be helped by drilling holes and filling them with fertile soil; the resulting improvement in deep irrigation and aeration increases root activity and produces healthy regrowth on old golf and bowling greens. In fact, such a regenerated green may be preferable to a new one because its compacted part provides a rigid framework to support foot traffic, while the deeper refilled areas are left free of compaction and provide for good plant growth.

Gardens, too, can have compacting problems, not

Costs of Agricultural Land," in *The Cropland Crisis: Myth or Reality?* ed. Pierre R. Crosson (Baltimore, Md.: Johns Hopkins University Press, 1982), pp. 188–89.

9. K. K. Barnes, H. M. Taylor, R. I. Throckmorton, G. E. Vanden Berg, and W. M. Carleton, eds., *Compaction of Agricultural Soils* (Saint Joseph, Mich.: American Society of Agricultural Engineers, 1971).

only from foot traffic but from the mechanized forms of tillage used by many gardeners. Rotary tillers, which combine the work of a plow, disk, and harrow in one operation and make a seedbed in one pass, have long been a mainstay of mechanized gardening. Dr. Konrad von Meyenberg, an agricultural engineer, was watching a dog dig at the edge of a flowerbed in Zurich in 1910. He noticed that the dog was not actually digging but scratching and scraping at the earth, with the result that the soil was broken up and turned over. Inspired by his observation, von Meyenberg designed and built the first rotary tiller, with tines that resembled claws. Experiments have shown that rotary tillers with L-shaped tines beat the ground at the bottom of their stroke and can produce a compacted soil layer. If the bottom of the L is removed from the tine, the soil is cut with a straight knife effect and plow pan formation is avoided.

Is there a chance that plows, disks, and harrows will entirely disappear from U.S. agriculture in the future? This is very unlikely, because some places are poorly suited to no-till agriculture. For example, in some poorly drained soils of cool climates, a cover of mulch would conserve too much moisture, which in turn would slow the warming of the soil and the rate at which crops sprout and grow.

Problems of Non-Tillage Systems

Non-tillage systems are not without some problems of their own. The mulch cover may serve as a habitat for insects, slugs, or rodent pests, and the farmer must identify problems of this sort and take appropriate pest control measures before the damage becomes serious.

Will the extensive treatments of soils with herbicides on unplowed farms create a threat to the environment? Despite the horror stories about some herbicides in the news media, well over 100 selective herbicides have been developed. Tests have indicated that if used properly, most of these have relatively low toxicity to mammals, other wildlife, and fish. Nor are herbicides a major hazard to the soil microbes; in fact, nearly all are organic molecules that serve as food for microbes. The fact that sensitive plants can frequently be grown a few weeks or months after the application of herbicides demonstrates the voracious appetite of microbes for even these weed killers.

Why don't the herbicides kill soil microbes as they kill weeds? Herbicide toxicity to weeds is generally achieved by such mechanisms as blockage of the photosynthetic, food-producing process of green plants or disruption caused by simulated growth hormones. But the decay microbes do not carry out photosynthesis or have the same type of growth hormones as weeds. The herbicides may have other metabolic effects on the microbes, but these are minimized by the low concentration of herbicides needed for weed control, which is typically equivalent to only a few parts per million of soil.

Even though herbicides do not appear to be leading to environmental disaster, it would be going too far to give them a totally clean bill of health. Scientific evidence gathered to date about the impact of herbicides on beneficial plant and animal life have not supported the most dire of the warnings of some environmentalists, but future research may turn up some surprises.

Because farming without plowing dramatically reduces erosion and frequently reduces runoff, there will often be smaller amounts of herbicides and other pesticides carried—either linked to sediment particles or dissolved in water—to surface waters or adjacent untreated fields than from conventionally tilled fields that have been treated with pesticides. (Note: "Pesticide" is a general term that includes herbicides as well as in- secticides, fungicides, etc.) However, if very heavy rain —rather than better infiltrating moderate rain—falls first on recently applied herbicide that is still on the surface, large amounts of the chemicals may be carried by runoff to adjacent nontreated fields or to streams, lakes, and estuaries.

In some locations herbicides have been found in ground waters. Although their concentrations have generally not exceeded suggested health standards, and no verified cases of health problems appear to have been traced to drinking herbicide-contaminated water, there is clearly a need for concern. Care will be re- quired in the use of reduced tillage or non-tillage meth- ods in areas prone to water contamination.[10]

Continuing improvements in weed killers may bring greater peace of mind for many people. One promising new herbicide, called ALA (delta-aminolevulinic acid), was developed by a team of scientists at the University of Illinois and is made from an amino acid found in all plants and animals. The chemical is sprayed at dusk, absorbed by weeds at night, and converted by the doomed plants into substances that are extremely sen- sitive to light. After the sun rises, the light triggers a chemical reaction that ruptures plant cells. The weeds wilt and die within hours, and the lethal substances quickly break down in the soil. This weed killer may be on the market as early as 1987.

10. Council for Agricultural Science and Technology, *Agricul- ture and Ground Water Quality*, report no. 103 (Ames, Iowa: CAST, 1985).

Riders

8
The Gospel According to Saint Guy and the Sagebrush Rebellion

How good is your vocabulary? Have you ever heard of a Fluvaquentic Medihemist, a Typic Gibbsihumox, a Hapludic Vermiboroll, or an Orthoxic Quartzipsamment? These names, which sound like something from *Alice in Wonderland*, represent different types of soils, with each of the syllables actually describing some soil property. For example:

fluv = floodplain;
aquentic = wet, water-deposited;
medi = of temperate climates;
hem = intermediate stage of decomposition;
ist = bog soils.

Therefore, Fluvaquentic Medihemist soils are fairly decomposed, wet, bog soils from the floodplains of temperate regions.

Who invented this method of making up so many wild and crazy names for soil? The examples shown above are only a tiny fraction of the total; they come from *Soil Taxonomy*, a handbook of the method used for soil classification in the United States, which was devised by the late Guy D. Smith and other members of the Soil Survey Staff of the U.S. Department of Agriculture.[1] This is the "Gospel of Saint Guy."

1. Soil Survey Staff, *Soil Taxonomy: A Basic System of Soil*

Mapping Soils

The guidelines for soil classification in *Soil Taxonomy* serve as the basis for soil surveys. The soil mapper has no easy job. Many of the soil features used in classification are hidden underground, so pits about five feet deep and wide enough to move around in must be dug —usually by hand, because digging machinery is generally not available. Some layers are so hard that a jackhammer may be needed to break through them. Old soil exposures in road cuts or ditches are almost always unsuitable for soil characterization because features become changed by prolonged weathering.

After the hole is dug, the soil surveyor gets down into the pit to look for distinctive soil horizons. These more or less horizontal layers differ from one another in such features as clay content, color, amount of organic material, or kinds and amounts of various salts. The types and arrangement of horizons determine how a soil is classified. Sometimes the soil layers are clearly different from one another, but many times the differences are more obscure; conflicts can occur when experienced soil mappers, who usually work alone, are brought together to compare their interpretations of the soil horizons exposed in a soil pit. Six experienced soil surveyors were once asked to describe a well-drained soil and a poorly drained soil in an area where all had done mapping. When the descriptions were tabulated and the judgments compared, there were six firm, widely ranging, and differing opinions of the soil hori-

Classification for Making and Interpreting Soil Surveys, Agricultural Handbook 436 (Washington D.C.: USDA, 1975). Soil surveys, interpretations, and land use planning are discussed in R. L. Donahue, R. W. Miller, and J. C. Shickluna, *Soils: An Introduction to Soils and Plant Growth*, 5th ed. (Englewood Cliffs, N.J.: Prentice-Hall, 1983), pp. 554–73.

zons, depths, colors, and other features of the same soils.

Another problem is that the practical considerations of mapping huge areas permit study holes to be dug only at wide intervals; thus the properties of much of the mapped soil must be inferred from observation of an extremely limited part of the total. Experienced soil surveyors can become skilled at guessing what is below the surface by relating observations in soil pits to surface features and by collecting additional subsurface samples with a soil auger; still, some uncertainty inevitably exists about the soil maps.

After the field work, lab work, analysis, and writing are all done for a specified area, a report with maps and interpretations is prepared. You may wonder whether you can pick up a survey report from the Soil Conservation Service (the federal agency responsible for most soil surveys) and find out what kind of soil you have on your own property. Your chances of getting a detailed soil map are much better if you live in a rural location than in an urban setting because soil mapping has been done most intensively in agricultural areas. The surveys that have been completed for more widespread regions are general estimates that may not be correct for a specific site. Even when more detailed surveys are available, on-site investigations are needed for particular locations. Residential sites, for example, can lose their surface soil to grading, and then they no longer resemble the original soils mapped in those areas.

Despite the unavoidable uncertainties of soil surveys, however, the maps and associated interpretations are helpful guides for farmers, county planners, civil engineers, and others. Useful nonfarm interpretations in soil surveys include ratings of soil suitability for such uses as: (1) sites for residential, industrial, or recreational development; (2) routes for roads, pipelines, or

underground cables; (3) sources for building materials
—gravel, sand, stone, or clay; (4) sources of topsoil to
be removed and sold for use in gardens; (5) potential for
sewage lagoons or septic tank drain fields.

Soils for Burying Waste

One of the most important nonagricultural uses evaluated in soil survey reports is how well suited different soils are for burial of solid wastes—the most common method of municipal waste disposal. These areas
of soil-covered garbage are called "sanitary landfills,"
but they are not always so sanitary. Noxious waste materials—medicines, pesticides, solvents, toxic chemical cleaners, sewage sludge—may pollute ground waters, and gases escaping from the rubbish can pollute
the air.

If the soil at the landfill site is very permeable or
close to the ground water, a bottom-and-side sealing
barrier of clay, soil cement, asphalt, or plastic liners
may be needed to control leaching. Those same lining
materials can also trap gases within a landfill; therefore, perforated suction pipe may be installed in gravel-
filled trenches to remove gas before high concentrations build up in the trash.

Why should there be concern about a buildup of gas
in the garbage? Don't dump sites always stink anyway?
The problem is not just offensive odors—the gas can
lead to catastrophic explosions. In 1977, workmen were
building a five-foot-diameter conduit to carry water to
the eastern part of Denver, Colorado. During the final
stages of inspection and cleaning before the water was
to be turned on, a welder asked two men to put in a fan
at a distant manhole. One of the workmen lit a match at
the manhole and all hell broke loose: the blast shot
flames 40 feet into the air; the fire created a vacuum

that pulled more oxygen into the conduit; four more explosions in the next 90 seconds blew the welder about 35 feet from where he had been standing; two workmen were killed and four firemen were taken to the hospital to be treated for carbon monoxide inhalation.

What had caused the explosion? Methane gas, produced by bacteria that were decomposing organic matter in a nearby landfill, had diffused laterally through the soil and entered one end of the conduit. About 5 percent by volume of methane gas is the minimum needed for an explosion, and tests of gas collected at the end of the conduit showed measurements as high as 14.4 percent. In the landfill itself, measurements reached concentrations as great as 52.2 percent methane.

Bureaucrats and Sagebrush Rebels

While most soil surveys are helpful to the various people concerned with soil, one particular interpretation contributed mightily to the furious attack on federal land management known as the "Sagebrush Rebellion."

Cattlemen have long been accused, and often with reason, of overgrazing the public lands, with resulting destruction of natural vegetation and reduction in food for wildlife. The battle against overgrazing reached a peak in the mid-1970s. New environmental regulations and a successful lawsuit by a conservation organization, the Natural Resources Defense Council, pressured federal land managers in the U.S. Bureau of Land Management (BLM)—an agency of the Department of the Interior—to adopt a new system designed to predict more accurately how much grazing should be allowed.

The method for evaluating grazing capacity was called the *soil-vegetation inventory method* (SVIM) and was set up as follows:

1. It was assumed that in areas with a similar environment (closely related soil and climatic conditions—called a "range site") there would be similar compositions of plant species and nearly equal amounts of edible forage production per acre.

2. For each type of range site, the government people sampled vegetation in an undisturbed location to determine the "standard" plant species mix and the "standard" average yearly amount of edible forage production per acre.

3. If the plant species composition or forage production in any area differed appreciably from the "standard" values expected for that type of range site, it was assumed that the difference was caused by overgrazing.

4. The U.S. government land managers used the figures of edible forage production on range sites to estimate how many cattle could be kept on the ranchers' grazing leases. Some severe cuts were ordered on lands rated as overgrazed.

As the civil servants responsible for its adoption should have realized immediately, the SVIM plan for calculating grazing levels was completely inaccurate. It is virtually impossible to estimate how much edible forage will be available on semiarid western rangeland for the obvious reason that rainfall frequently varies greatly from year to year. A wet year may produce two or three times more forage from woody plants and ten times more from annual plants than a dry year, and there is no way to predict how wet the next year will be. The only way to manage such a range without overgrazing, then, is to keep the livestock numbers at a consistently low level, so that even in dry years there will be enough forage. An alternative is to permit more animals to be added during the course of a good year, but such flexibility creates its own difficulty: ranchers prefer a guaranteed number of livestock, to give their leases known value in case that lease is sold or used to obtain a loan. For this reason, the BLM now grants ten-year grazing leases with a specified number of animals.

How can you tell whether an area is being overgrazed? The SVIM plan of comparing the observed species composition on a range site to the expected "standard" species mix can be very misleading because chance plays a large role in plant distribution; thus the differences may be natural and not the result of overgrazing. The evaluation of grazing range is often more of an *art* than a *science*, but an experienced range specialist can generally tell at a glance if the vegetation on a range is being eaten at a much faster rate than it can grow. The most accurate way to determine whether a rancher's livestock numbers are about right is to fence cattle out of an area of range and then compare the protected vegetation with the adjacent grazed vegetation over a period of several years.

Soil-Vegetation Inventory Method in Action

What happened to the SVIM? Shortly after adopting the new system, the BLM heaped lavish praise on its own plan, exuberantly predicting that the following benefits would result from the use of SVIM procedures:

> The annual harvest of meat, wool, timber products, pinyon pine nuts, and other products will never exceed the annual allowable productive capacity of any parcel of public lands. Kids will brag about their dads and moms working for the B.L.M. B.L.M. employees will proudly proclaim in a crowded pool hall in a small ranch town that they

work for the B.L.M. Fenceline contrasts will have disappeared between ownership boundaries because everyone will be following the good example of the B.L.M. The state legislatures will be asking the B.L.M. to take over the management of state-owned land.[2]

The cattlemen were far less enthusiastic, and soon many of them passionately embraced the aims of the Sagebrush Rebellion—to prevent cuts in their grazing privileges. The rebellion was led by Nevada livestock raisers who were threatened with large financial losses: if the BLM declared their leased public lands overgrazed and reduced their livestock grazing allotments, the leases would have far less value if they were sold to other operators.

On the Fourth of July, 1980, a crowd of people from Moab, Utah, stood cheering east of their town as the county roads superintendent, in a well-publicized show of defiance toward federal attempts to more closely regulate grazing and mining on public lands, bulldozed through federal lands under consideration for wilderness protection. Some local residents described their Independence Day demonstration as the "first shot of Utah's Sagebrush Rebellion."[3] Six western states— Nevada, Arizona, New Mexico, Utah, Washington, and Wyoming—passed laws claiming about 114 million acres of federal lands within their borders. In total the sagebrush rebels proposed transferring about 400 million acres of public federal land to the states. Nearly all of this public land, now managed by the Bureau of Land Management is in the western United States.

2. U.S. Bureau of Land Management, *The Soil Survey Program of the Bureau of Land Management, Past, Present, and Future* (Washington, D.C.: BLM, 1980), p. 195.
3. William Boly, "The Sagebrush Rebels," *New West* 5, no. 22 (November 3, 1980): 19.

The Sagebrush Rebellion ended not with the transfer of federal lands to the states but with the election of Ronald Reagan, whose administration enacted policies on federal land use more in line with the desires of the sagebrush rebels.

The BLM pushed full speed ahead with the SVIM, hiring many new employees to carry out the inventories. Many other employees criticized the grazing evaluations as worthless. I wrote several memoranda myself, pointing out inaccuracies, and sent a copy of one memo to the inspector general's office in the Department of the Interior. In 1982, the SVIM was shown to be grossly inaccurate by a National Academy of Sciences report, *Developing Strategies for Rangeland Management*. Dr. John E. Menke, one of the contributing authors, concluded that "the often semi-arid nature of our rangelands makes reliable estimation of carrying (grazing) capacity virtually impossible."[4] The BLM's enthusiasm was soon over: an instruction memorandum in December, 1982, declared that "the S.V.I.M. is no longer an approved inventory method."[5]

4. John W. Menke and Michael F. Miller, "Sampling and Statistical Considerations in Range Resource Inventories," in National Research Council, National Academy of Sciences, *Developing Strategies for Rangeland Management* (Boulder, Colo.: Westview Press, 1984), p. 805.

5. United States Department of the Interior, Bureau of Land Management, Instruction Memorandum No. 83-155, Subject: SVIM Program and Archiving of Data, December 2, 1982.

Desert Dunes

9
The Desert — Delicate or Durable?

Nature's Paving

Armor, a vital source of protection in helmets, warships, airplanes, and fortifications, also plays an essential part in protecting some areas of desert soils. The "desert armor," also called "desert pavement," is composed not of sheets of metal but of gravels and stones that protect underlying soil from the erosive forces of wind and water.

Often acres in extent and essentially barren of vegetation, desert pavement looks as though countless pebbles and small rocks had been spread over the ground surface, rolled flat with a roller, and covered with oil.

How did these strange-looking surfaces form? Apparently the gravel and rocks became concentrated at the surface after wind or runoff from rainfall removed finer soil particles, leaving a residuum of coarse surface material. Rocks originally buried in the soil may also have been lifted by such processes as wetting and drying of the soil. When clays—common in pavement soils—become wet, they swell and push rocks upward. After drying, the stones often stay in their uplifted position, and soil moves into cracks beneath the rocks. Such a process repeated many times may move buried rocks to the surface.

The oillike covering of desert pavement is called "desert varnish." A popular current theory is that it is formed by the gradual addition to rock surfaces of windblown dust rich in iron, manganese, and clay. It has probably taken thousands of years for the varnish to form on the pavement.[1]

Dunes

In strong contrast to the soil protected by desert armor are the loose, constantly shifting sands that form dunes. Most dunes are composed of quartz (silicon dioxide) grains; when examined under a hand lens or low-powered microscope, these sand-blasted particles can be seen to have round or oval shapes and frosted surfaces. Not all dunes are made of quartz, though. New Mexico's snowlike dunes at White Sands National Monument are composed almost entirely of gypsum, the material used to make plaster of paris; dunes in Bermuda and Hawaii are made of calcite (calcium carbonate), the smoothed and eroded skeletal remains of tiny marine animals; and sections of dunes along the Gulf Coast are made of clay.

Dunes are among the most attractive of desert lands. As the late desert naturalist Edmund C. Jaeger pointed out in his book *The California Deserts*:

> Any land area dominated by dunes should be often visited, if for no other reason than to study light effects and glorious colors. Every hour of the day marks changes, and morning or evening the clean sands, viewed from a distance, are marvel-

1. For further discussion, see Christopher D. Elvidge and Richard M. Iverson, "Regeneration of Desert Pavement and Varnish," in *Environmental Effects of Off-Road Vehicles*, ed. R. H. Webb and H. G. Wilshire (New York: Springer Verlag, 1983), pp. 225–43.

ously rich in translucent shades of purple, pink, blue, or yellow. Partly clouded, moonlit skies add to the effect, and under the clearing skies following rains every dune is a changing panorama of glory.[2]

The pioneering studies about sand dune formation and movement were done by Ralph Bagnold, a British army officer, explorer, and scientist. While stationed in Cairo in the late 1920s and early 1930s, Bagnold explored the dunes of the Great Sand Sea in the Sahara, traveling in the precursor of today's dune buggies—the Model T Ford. The Ford Motor Company claimed for the Model T the slogan, "No hill too steep, no sand too deep," and Bagnold and his companions found that the company's pledge was accurate.[3] Bagnold's *Libyan Sands* describes his first trip up a dune:

A huge glaring wall of yellow shot up high in the sky a yard in front of us. The lorry [Model T] tipped violently backwards—and we rose as in a lift, smoothly without vibration—. We floated up and up on a yellow cloud. All the accustomed car movements had ceased; only the speedometer told us we were still moving fast. It was incredible. Instead of sticking deep in loose sand at the bottom as instinct and experience both foretold, we were now near the top a hundred feet above the ground.[4]

Bagnold discovered the reason he could so easily drive up the dunes. Sand blows up the windward face

2. Edmund C. Jaeger, *The California Deserts*, 4th ed. (Stanford, Calif.: Stanford University Press, 1965), p. 29.

3. Burkley Belser, "Bagnold of the Dunes," *Science 82* 3, no. 2 (March, 1982): 31.

4. Ralph A. Bagnold, *Libyan Sands* (London: Hodder & Stoughton, 1935), p. 129.

of a dune, forming a slope of 10 to 20 degrees, and after moving sand reaches the crest, it avalanches down the lee side, forming a steeper slope of around 35 degrees, called the "slip face." Grains of sand blown upward toward the dune crest are packed closely and thus provided enough support for the Model T to reach the top. The sand pouring down the slip face, in contrast, forms loose sandpiles, but the momentum of the car carried it over the downwind slope where, in places, areas of sand were "so soft one could plunge a six-foot rod vertically into them without effort."[5]

Bagnold and his companions explored the previously unknown sand dune territory in three Model Ts, carrying extra food, water, spare parts, and gas, and using overflow cans to retain water splashed from boiling radiators. Even severe mechanical car trouble was not disastrous, since they had two extra vehicles. During World War II, at Bagnold's initiative, a small British mobile striking force used his knowledge of dune travel to harass Italian troops in Libya.

Bagnold had returned to England in 1935 to make laboratory studies of sand movement, using a wind tunnel. There he discovered that sand moves downwind in two steps. In the first, called "saltation," sand is carried into the air by the wind and then pulled back to earth by gravity. The falling grains crash into grains on the surface, knocking them forward like balls in a game of billiards. As the bouncing and rolling grains fall over the crest, the dune creeps forward a fraction of an inch. As long as the amount of sand blown onto the windward slope is greater than the amount blown over the crest, the dune grows higher. Eventually dunes stop growing, but they may keep moving downwind several feet or more a year. One dune in Asia reportedly moved 60 feet in one day.

5. Ibid., p. 131.

The Desert—Delicate or Durable?

The inexorable movement of dunes may cover buildings or towns. At Cape Henlopen, Delaware, dunes traveling 60 feet a year swallowed up a historic lighthouse and a nearby forest. Fifty years later the dunes had moved on, reexposing a number of buildings that remained well-preserved. Other wind-driven dunes, reaching heights of 250 feet, gradually buried several small coastal towns along Lake Michigan.

The Bedouin protect their water holes from creeping sand dunes by placing stones or pebbles on dune crests. Sand grains rebound more easily off the hard rock fragments than they do off sand, and the grains striking the stones consequently bound away farther and faster than sand on sand, and so the dune dissipates.

Because the Bedouin technique was not sufficient to protect oil rigs from encroaching sand dunes, Bagnold was hired by the Shell Oil Company in 1946 to find a method of keeping the sand away from oil fields. The solution was very simple: building fences upwind restrains the sand; as sand collects, the fences are built higher. In Kuwait, the largest oil deposit in the world is protected by six miles of such fences.

Another phenomenon Bagnold investigated was that of the "singing sands." Wind blowing over sand dunes may produce an eerie sound, loud enough to make conversation difficult. This strange dune music may have been responsible for Saharan legends about waterless sirens and lost monasteries with bells ringing. In his laboratory Bagnold was able to measure the frequency that creates singing sands and to duplicate the peculiar phenomenon.

Off-Roaders vs. Environmentalists

Improvements in off-road vehicles since the Model T have made desert driving a popular sport. Several hundred thousand people a year may be involved in off-road vehicle travel in the deserts of southern California. Has such traffic become so intense that desert soils are being severely damaged? The sand in the dunes is little changed by driving, although plant and animal life may be run over and destroyed. The soils on the playas (clay-rich dry lakes) most favored by off-road drivers—those with hard, dry surfaces and limited numbers of mud cracks—are also resistant to harm by vehicles. But most soils in the California deserts are not in sand dunes or playas, and can indeed be damaged. Soils on steep motorcycle hill climbs, for instance, are displaced downslope by tire action; when strong rains come, the runoff cuts deep gullies in these smoothed, loosened, precipitous trails.

Compacted tracks made on level areas can cause reductions in plant growth for many years. Other scientists and I conducted experiments to determine the minimum amounts of driving and soil compaction necessary to reduce growth of desert annuals in following years. We found that as little as one pass of a truck on wet soil can reduce the cover of desert annuals in the track more than a year later.

A truck passing over wet soil causes the soil particles to be squeezed into a tightly packed grouping. When plants begin to grow in this compacted soil, its greater

density slows down root penetration and thereby re-
stricts the plant's access to water and nutrients. In contrast, a lighter vehicle such as a motorcycle, or a truck
driven on less easily deformed dry soil, will not notice-
ably compact soil until repeated passes have been made
over the same track. Many tracks made by a single pass
will be erased in a relatively short time by desert rain
and windstorms, but on the unique and interesting
surface called "desert pavement," even a single vehicle
pass can create a scar that may be visible for decades
or centuries.[6]

The use of public lands for off-road driving had led
to a storm of controversy between environmentalists
and recreational drivers. As the dispute about the
threat to the desert became louder and louder, Congress decided that the desert was indeed in danger and
went into action. Legislation was passed in 1976 that required the Bureau of Land Management to study the
resources needing protection in the desert and then to
develop and put into effect a plan to protect California's
public desert lands.

The huge size of the California desert (about as big
as the entire state of Kentucky) made it very difficult to
measure the total damage caused by off-road recreational vehicles (ORVs), but it was possible to measure
very heavily damaged areas by means of aerial photographs. About 41,000 acres of severely damaged land
was found in the massively tracked areas that radiate
from the recreationists' favorite gathering places. In
these innumerable intensely compacted trails, with
countless less-compacted tracks on the remaining surface, the driving has obliterated most of the smaller

6. See J. A. Adams, L. H. Stolzy, A. S. Endo, P. G. Rowlands,
and H. B. Johnson, "Desert Soil Compaction Reduces Annual
Plant Cover," *California Agriculture* 36, nos. 9, 10 (September-
October, 1982): 6–7.

shrubs, although many of the larger shrubs—which can easily puncture tires—have held their own.

Campsites and parking areas created and used by the off-roaders added up to some 6,000 acres more. These are stripped of vegetation and packed so hard that plants would have great difficulty returning to those locations. Hill climbs, where motorcycles are driven straight up very steep slopes, accounted for another 1,200 acres of devastated land.

Thus the total for the most ravaged areas of off-road driving as it stood in 1978 was at least 48,200 acres. This is about one part in 519 of the desert. How do these figures compare with demolishment of natural desert habitats by other types of disturbances? Paved and dirt road surfaces were calculated to cover some 96,000 acres in the desert, about twice the area estimated to have been very heavily affected by off-road driving. Communities in the desert were found to cover about 94,000 acres, similar to the area covered by roads. Desert farms had removed 870,000 acres of natural habitat, about 18 times as much as ORVs. Should we then conclude that damage caused by motorcycles and four-wheel-vehicles in the California desert is insignificant and that restrictions on this activity are unnecessary?

To answer this question, let us first visualize how large an area has been demolished. If the 48,200 acres of the most intense ORV damage could be gathered together and moved into a half-mile-wide strip with a road running through the center, it would take two hours and 44 minutes to reach the end of the strip at the legal 55 MPH speed limit. The driver would be staggered to see the amount of devastation.

Keep in mind that the calculated 48,200 acres of severe damage is an underestimate because some damage was undoubtedly overlooked, even though the on-the-ground experience of BLM rangers and recreation specialists was used to check the estimates made from aerial photos.

More important, large-scale off-road recreational driving, which began in the late 1960s, creates steadily increasing damage. Most of the aerial photos used in the survey were dated 1977 and 1978; some were taken as long ago as 1973 and 1974. Dove Springs Canyon, a heavily used hill-climb area, is one of the few sites where the rate of increase in ORV damage can be measured for recent years. Aerial photographs taken in 1982 show a 76 percent increase in severely damaged space (camping and parking areas, hill climbs, and dense trail coverage) compared with aerial pictures taken in 1973.

In addition to the desertwide total of more than 48,000 severely ravaged acres, there are huge areas of less intense off-road use where the destruction does not show up clearly in aerial photography but is still visible on the ground.

How long will off-road tracks last on desert surfaces? On sand dunes, where the loose sand particles are very resistant to compaction, tracks are soon swept away by the winds as the sand shifts. Beaten trails, however, remain visible for long stretches of time on most desert soils, as can be vividly seen at the old Desert Training Center, where military maneuvers were conducted under the direction of Gen. George S. Patton more than four decades ago.

The Desert Training Center was created in 1942 to provide the nucleus of an American "Panzer Army" in World War II. Patton personally chose the maneuver area—180 miles long and 90 miles wide in parts of California, Nevada, and Arizona—by flying, driving, and marching over the region accompanied only by a lieutenant colonel. General Patton made the survey under simulated combat conditions, allowing himself and his companion minimum rations, including one gallon of water a day apiece. Patton chose the area because in all that vast space, he and his aide did not meet another person, and because the contours of the terrain allowed opposing troops to make marches of up to 400 miles without seeing each other.

The troops, planes, tanks, and shelling are long gone, and the old military maneuver scene is again empty and quiet, being farther from the major population centers than the locations favored by the recreationists. Only a few of the structures remain: the back wall of a chapel, aircraft runways, and a large earthen topographic replica that was used by officers to plan maneuvers. But immense numbers of old roadways, tank tracks, and disturbed areas once occupied by tents are still very visible; the many intervening seasons with their driving winds and rain have failed to erase the evidence of past occupancy. The soil is still intensely compacted in areas that had heavy use, and the cover of vegetation is still severely reduced compared with nearby undisturbed locations.

Can the Bureau of Land Management, which was given the responsibility of developing a plan to control off-road destruction, confine the activity to zones that have already been severely damaged or that have little value biologically, aesthetically, or archeologically? In 1973 the BLM implemented the Interim Critical Management Plan and in 1980 completed the Desert Plan to restrict off-road activity to appropriate locations. The job may be impossible, however, with only 19 rangers to cover the 12 million acres of public land administered by the BLM in the California deserts. Indeed, the 1978 survey of severely damaged desert land showed 16,200 acres (hill climbs, campsites, and regions with a very high density of trails) in the BLM's open areas where off-road driving is permitted, but 32,000 equally ravaged acres in the "restricted areas" where driving was supposedly not allowed.

Valley Fever

Just as off-road drivers may damage desert soils, desert soils—through the spread of valley fever (also called "desert fever") spores—may damage off-road drivers.[7] Dr. Chester R. Leathers, a microbiologist at Arizona State University at Tempe, who specializes in the study of valley fever, isolated the fungus in one of the most intensely used off-road recreational areas in the California desert—Johnson Valley.

In the soil the valley fever fungi grow as fine, threadlike structures that produce hard-shelled, easily airborne spores. Once inhaled, these spores burst, releasing smaller spores in the body. The organism is found in very localized areas in the California deserts and San Joaquin Valley, and in very confined locations elsewhere in the southwestern United States, Mexico, Central America, and South America. Many people in those areas have had the disease without knowing it: skin tests indicate that over half of the people in California have been infected. About 60 percent of victims develop lasting immunity to the fungus without becoming sick at all. In the remaining 40 percent, symptoms range from a mild respiratory infection often mistaken for a cold to a lingering illness closely resembling pneumonia or tuberculosis. In a small number of cases the fungus may be fatal.

Potential victims include not only off-road vehicle operators but anyone in the area who disturbs the soil

7. Valley fever is covered in detail in C. W. Emmons, C. H. Binford, J. P. Utz, and K. J. Kwon-Chung, *Medical Mycology*, 3rd ed. (Philadelphia: Lea & Febiger, 1977), pp. 230–53.

and breathes the resulting dust: archeologists sifting dirt for artifacts, military personnel training for desert warfare, or a camper covering a campfire with earth. In December, 1977, a fierce dust storm in the San Joaquin Valley transported huge amounts of dust as far north as Sacramento, California, 300 miles away, carrying valley fever spores and causing hundreds of people to become ill.

And it is not only humans who suffer from valley fever. Mbongo, a gorilla at the San Diego Zoo, died from the disease. He liked to throw hay in the air and let it fall on his head, and doctors believe that he may have inhaled the fatal spores from the hay.

There is no known cure for the disease at present, and the most effective treatment is a powerful antibiotic—which often destroys the kidneys before helping the body overcome the disease. A vaccine is currently being tested; if it proves successful, it will be very useful for military personnel or other newcomers to an endemic area. For the first time, Americans may need shots just to move across the country.

Smoke

126

10
Sour Soils and Acid Rains

A Pioneer of Soil Chemistry

In November, 1860, Edmund Ruffin, a stern-faced Virginian with shoulder-length white hair, boarded a train for South Carolina. The 66-year-old Southerner showed his hatred for the North by brandishing a weapon symbolic of the abolitionists—one of John Brown's pikes, which Ruffin had obtained at Brown's execution a year before. "As old as I am I have come here to join you in (the) lead," he told his friends in South Carolina.[1] The Confederate forces arranged as a special compliment to the venerated Ruffin that he would fire the first shot against Fort Sumter. At dawn on a Monday, the 12th day of April, 1861, Ruffin fired the initial cannonball that smashed into the fort. A deafening roar of continuous firing followed, and the Civil War had begun. Ruffin proved one of the most implacable of the Southern "fire-eaters"; he committed suicide after Lee's surrender to avoid living under the "perfidious Yankee race."[2]

1. Horace Greeley, *The American Conflict* (Hartford, Conn.: O. D. Case, 1865), I, 335.
2. Avery Craven, *Edmund Ruffin Southerner: A Study in Secession* (New York: D. A. Appleton, 1973), p. 171.

Edmund Ruffin, old Secessionist,
Firer of the first gun that rang against Sumter,
Walks in his garden now, in the evening-cool,
With a red, barred flag slung stiffly over one arm
And a silver-butted pistol in his right hand.
He has just heard of Lee's surrender and Rich-
  mond's fall
And his face is marble over his high black stock.
For a moment he walks there, smelling the scents
  of Spring,
A gentleman taking his ease, while the sun sinks
  down.
Now it is well-nigh sunken. He smiles with the
  close,
Dry smile of age. It is time. He unfolds the flag,
Cloaks it around his shoulders with neat, swift
  hands,
Cocks the pistol and points it straight at his heart.
The hammer falls, the dead man slumps to the
  ground.
The blood spurts out in the last light of the sun
Staining the red of the flag with more transient
  red.[3]

Ruffin's suicide ended a life remarkable for much
more than his part in the rebellion. He was also a scientist, publisher, and successful farmer. In his earlier
years of farming, Ruffin had been very unsuccessful,
struggling to avoid losing his land near Petersburg, Virginia. When his studies of the fertile limestone soils
of the Shenandoah Valley showed that they had some
acid-soluble calcium, whereas his did not, he applied
calcium to his own infertile fields in the form of
crushed oyster shells.

3. Stephen Vincent Benét, *John Brown's Body* (Garden City,
N.Y.: Doubleday, Doran, 1928), p. 370.

Ruffin kept careful records of the amounts of shell used and the crop yields before and after the additions. He learned that the oyster shells increased crop yields, and thus he became the first American farmer to add lime for the purpose of neutralizing soil acidity. Ruffin discovered that too much lime could be harmful as well; he accurately described zinc deficiency in over-limed corn. Ruffin's book, *An Essay on Calcareous Manures* (1832), was one of the first scientific studies of soils and crops.[4] He has been called the father of soil chemistry in America.

What caused Ruffin's soils to be so acid, and why did the addition of oyster shells increase the fertility of his fields? Before we answer this question, it will be necessary to look at a very brief and simple explanation of acidity and consider how acids can be neutralized.

Soil Acids

The French chemist Antoine Laurent Lavoisier (1743–94) thought that oxygen was an essential component of all acids; he devised the name "oxygen" from Greek words meaning "acid producer." But Lavoisier was wrong: hydrogen, rather than oxygen, is the key atom in acids.

When certain molecules are dissolved in water, some of their atoms break away. These "dissociated" atoms are not ordinary atoms, but electrically charged "ions" (from a Greek word meaning "wanderer"). Acids release hydrogen ions that produce a characteristic sour taste when safely diluted. In contrast to the acids, another group of substances, called "bases" (strong bases are termed "alkalis"), taste bitter and can neutral-

4. Edmund Ruffin, *An Essay on Calcareous Manures* (Petersburg, Va.: J. W. Campbell Publishing, 1832; reprint, John Howard Library, 1961).

ize acids. Calcium, magnesium, potassium, and sodium have the ability to form bases in the soil after weathering releases them from minerals. Soils that have not been highly leached, such as many of those found in arid or semiarid parts of the western United States, tend to retain large amounts of these substances and so are typically alkaline rather than acid. Conversely, soils in areas of higher precipitation (in general, permeable soils with an average of more than 20 inches of rainfall per year) eventually lose a large part of these acid-neutralizing substances by leaching and therefore become acid.[5]

Soil acids come from a number of sources, including carbon dioxide gas, which dissolves in water to form carbonic acid; organic matter, which decomposes to form organic acids; certain types of acid-forming fertilizers; and human-caused acid rain (which is discussed later in this chapter).

The intensely leached soils on Ruffin's Virginia farm had lost large parts of their alkalinity-producing elements and so had become "sour" soils with a predominance of soil acids. When Ruffin applied oyster shells, the calcium from the shells acted to neutralize soil acids.

There is a question whether it is the hydrogen ions in acid soils that suppress the growth of acid-sensitive plants. You might imagine that soil water could become acid enough to burn the plant roots, just as concentrated sulfuric acid can burn clothes and skin, or even blind a person. Actually, however, even the most strongly acid soil water ever encountered (and it is formed only rarely in organic soil surface layers) is less acid than vinegar, grapefruit, or even apples. The more

5. Soil acidity and alkalinity are covered in N. C. Brady, et al., *The Nature and Properties of Soils*, 8th ed. (New York: Macmillan, 1974), pp. 372–403.

typical soils of humid areas are less acid than tomatoes or beer—about as acid as green beans. Therefore the acidity in soils is not strong enough to burn plants, and studies by scientific investigators have shown that plant damage in even strongly acid soils is not caused by the acidity itself. What, then, causes the damage to acid-sensitive plants?

The Role of Aluminum in Acid Soils

Dissolved aluminum is the most common cause of crop failures in acid soils. Aluminum is the most plentiful metallic element in the earth's crust; only the nonmetals oxygen and silicon are found in greater abundance. Although aluminum is more common than iron, it is locked up in very stable compounds. Iron has been known and prepared from its ores since prehistoric times, but aluminum was not even recognized as a metal until 1827, and not until 1855 was a method devised to obtain reasonably pure aluminum in moderate quantities. The metal remained expensive, being used for ostentation, such as for the rattle of Napoleon III's infant son or the cap at the top of the Washington monument. Finally, in 1886, Charles Martin Hall, a young American, devised a cheap method for extracting aluminum from its ores; this light, malleable metal could then be used for even the most common purposes, such as today's kitchen utensils and the ubiquitous beer can.

Although humankind was slow to discover aluminum and extract it for widespread use, soil acids through the ages have dissolved it, and in strongly acid soils the aluminum in the soil water may be extremely toxic to some species of plants. Moreover, the dissolved aluminum in the soil water reacts with water molecules to produce additional acidity, which in turn can dissolve even more aluminum. Thus dissolved aluminum is a cause of soil acidity as well as a result of it.

Acid Rain

The acidification of soils and waters by acid rain and snow is widely regarded as one of our most severe environmental problems. Sulfur oxides and nitrogen oxides—gases released into the air by burning fossil fuels—are changed by chemical reactions in the atmosphere into sulfuric acid and nitric acid, respectively. These acids return to earth in rain, which has raised fears that a gradual acidification and sterilization of many soils and waters may occur, particularly in southern Scandinavia, southeastern Canada, northern New England, and parts of the Adirondacks. Aluminum and other metals dissolved by acids can be toxic to fish as well as plants.

Despite the strong concern that has developed in recent years, acid rain is undoubtedly not a new phenomenon. In 1799, Joseph Black, the discoverer of carbon dioxide, died and was buried in Edinburgh. In less than 80 years the inscription on his marble tombstone had become illegible. A search for the culprit revealed that part of the erasure was accomplished by the gas he discovered, carbon dioxide, which dissolves in water to form a weak acid called carbonic acid. In addition, smoke from burning coal may have led to the formation of rain containing the stronger sulfuric acid (which would have significantly increased the rate of weathering); if so, this was the acid rain—rain with more acidity than that caused solely by carbonic acid—of the nineteenth century. Acid-forming substances may also have moved directly from smokestacks to the moistened tombstone surface without being carried down in rainfall. Another defacer of Black's tombstone was frost, which pried rock fragments loose where water froze and expanded between grains.

On the average, about a third of an inch has been weathered away from limestone and marble in Edinburgh every 100 years. Tombstones made of slate, which is more resistant to acid solutions, were barely roughened after a century in Edinburgh cemeteries. In small Scottish towns that have had less coal smoke and thus smaller amounts of acid-forming substances in the air, the limestone and marble are dissolving more slowly. Nevertheless, enormous numbers of buildings and monuments constructed of the materials are threatened with premature defacement by air pollution in Europe, North America, and elsewhere. The acid solutions can also react with such materials as bronze, paints, leather, textiles, and paper.

There are claims that many of Holland's 15,000 steeple and carillon bells are now out of tune because of acid rain. The environmental ministry of the Netherlands, which commissioned a study of this matter, believes that acid rain is thinning the metal walls of the bells and lowering their pitch. The corrosion affects small bells faster than large ones, lowering their pitch disproportionately and causing the carillons to become dissonant. The only remedy is to lower the pitch of the larger bells as well by carefully scraping their inside walls.

What is the difference between a weak (e.g., carbonic) acid and a strong (e.g., sulfuric) acid? When a strong acid reacts with water, it produces a high proportion of acidity because a large fraction of its hydrogen atoms are released as electrically charged ions. A weak acid produces a smaller proportion of acidity because it releases a relatively small number of hydrogen ions.

Although strong acids may cause harmful effects as

acid rain, they are among the most important chemicals we have. Their ability to dissolve substances and carry through chemical reactions makes them vital to innumerable industrial processes. The early Greeks and Romans were limited to weak acids—vinegar being the strongest. Strong mineral acids were discovered by the medieval alchemists in their attempts to turn common metals into gold. Ironically, if they had succeeded in manufacturing gold, its value would have disappeared as soon as it was no longer rare; they did not know that the strong mineral acids would prove a far more valuable discovery, one that actually increased in importance as the acids became cheaper and more plentiful.

Is acid precipitation caused only by human activity? Acid rain has been measured in very remote locations, far from industrial sources, including weather stations in Venezuela, Bermuda, an island in the Indian Ocean, and Katherine, Australia. The composition of the rain showed some evidence of human influences, but the measurements indicate that natural sources of acid rain may also be significant. Relatively large amounts of acidity in Antarctic ice core samples up to 350 years old also give evidence of natural acid precipitation. Acid-forming gases can come from such varied natural sources as ocean spray, volcanoes, decaying organic matter, and lightning as well as from the acidifying substances emitted by smokestacks and automobile exhaust. The importance of natural sources in contributing to acid rainfall has not yet been clearly determined.

The strong (sulfuric and nitric) acids in rainwater are far from concentrated. The acid rains will not burn the plants or animals on which they fall because, to use a familiar comparison again, the acidity of even the most strongly acid rains rarely exceeds that of apples.

Moreover, the acid precipitation in the northeastern

United States, southeastern Canada, and Scandinavia is falling on soils that generally have already been made very acid by natural soil formation processes. Some of these soils are among the most acid in the world, particularly those commonly found under coniferous forests in cool, humid climates, and the acid peats. The results of natural soil formation in these humid areas are the same as those attributed to acid rain: acidification of soil and water, leaching of nutrients, and release of aluminum.

Buffered Soils

It seems logical to think that the acid rains will soon make these acid soils much more acid than they already are, but in fact, the soils are surprisingly resistant to further increases in acidity. Much of the acidity in the strongly acid soils results from organic acids created by the decomposition of plant litter. These organic acids release hydrogen atoms as electrically charged ions, but, being weak acids, release only a small part of their potential acidity. When soil acidity is augmented by acid rains, the organic acids—in preserving chemical equilibrium—dissolve in even smaller-than-usual amounts; the result is little net change in acidity. Both organic matter and clays can act as buffers, storing extra acidity, because the surfaces of these materials attract added hydrogen ions from the soil water and prevent large changes in acidity.

The same buffering action, with its job of maintaining chemical equilibrium, causes the acidity held by soil humus and clay particles to resist attempts to make the soil water less acid. When you add lime, a pound or two should generally be enough to neutralize all the "active" acid content of the soil water in the top six inches of an acre of even strongly acid soil. However, the active acidity will then be replaced by reserve acidity, released by soil humus and clay particles, and this process will continue until all the reserve acidity is neutralized as well. A clay soil high in organic matter, for example, may have 50,000 or even 100,000 times more reserve acidity than active acidity. This is why you will most likely have to apply several tons of lime to neutralize a substantial part of the acidity in the top six inches of an acre of strongly acid soil.

To find the soil most susceptible to acidification by acid rains, you would look for the type composed mostly of coarse materials with relatively small amounts of soil acidity, clay, and organic matter. Little research has been done about the effects of acid rains on such soils.

Acidified Lakes

Just as soils are naturally acid in the humid areas where the greatest concerns have been raised about acid rain, runoff from rains in these areas will be acid regardless of the acidity of the rain. The amounts of acid in the soils are much greater than in the rains, and the runoff water from heavy rains and rapid snowmelts is acidified by contacts with acid soil layers. Therefore, where water drainage into lakes and streams reaches a chemical equilibrium with the soil during its passage, natural soil formation may often be more important than acid rain in determining the acidity of lakes and streams.

If you looked for conditions where acid rain would be the predominant element in acidifying a lake, you would find that this is most likely to happen where lakes have relatively little bicarbonate alkalinity (the principal measure of a lake's ability to neutralize acids), and where the size of a lake is large in proportion to the area it drains. In the latter situation, much of the acid precipitation entering the lake either falls directly into it or runs off adjacent areas of bedrock (minimizing contact with soil) or through shallow layers of coarse soil (which typically have little neutralizing capacity).

Acidification Caused by Changes in Land Use

Ed Krug and Charles Frink, soil scientist and vice-director, respectively, of the Connecticut Agricultural Experiment Station, suggested in an article for the journal *Science* that many soils and waters may be acidified not by acid rain but as a result of changing land use.[6]

By the early 1900s essentially all the spruce-fir forest lands of northern New England and on the steep slopes of the Adirondack mountains had been clear-cut for pulp, except for a small part of the forest in the Adirondacks that was preserved. The original forests were very resistant to fire, but lumbering was frequently followed by severe fires that caused the organic acids at the soil surface to be oxidized and lost in ash. Soil in these disturbed areas typically became much less acid than it had been before clear-cutting.

The forests, overexploited both in eastern North America and in Europe, are now returning to many of the formerly clear-cut areas. For example, in New England the volume of standing wood increased by about 70 percent between 1952 and 1976, and in southern Norway between 1927 and 1973 the volume of standing wood increased by an even greater proportion. Reestablishment of the forests creates thickened layers of plant litter on the soil, greater quantities of organic acids, and therefore more acid soils.

A Norwegian scientist named I. T. Rosenqvist illus-

6. E. C. Krug and C. R. Frink, "Acid Rain on Acid Soil: A New Perspective," *Science* 221 (August 5, 1983): 520–25.

trated the astounding increase of acidity that can come from reforestation. He determined that the reserve acidity in humus under a 90-year-old spruce forest on an abandoned Norwegian farm was equivalent to the strong acids that would be added in about 1,000 years of strongly acid rains falling at a rate of 40 inches per year.[7]

Because increasingly acid soils can produce increasingly acid runoff, changes in land use and subsequent reforestation may also be playing an important part in acidifying lakes.

Recently there have been reports that in some sections of forests in the eastern United States and West Germany, barren patches are forming where trees are losing their foliage and dying. In addition, many trees in these areas appear to be growing more slowly than they once did. Can this decline be attributed to increased acidification of soil by acid rains? As we discussed earlier, many highly acid soils in these areas have a strong capacity to resist increases in acidity, but some soils with small amounts of clay and organic matter are more poorly buffered against acidification from rains.

However, there are many other suspects for damage to the forest.

1. Lack of water: the decline in eastern U.S. forests began about the time of the great northeastern drought of the early 1960s.

2. Air pollution effects: (a) dry deposition—fallout of acids or acid-forming substances in solid particles, or absorption of acids or acid-forming substances in gaseous forms—may be a more serious problem than acid rain and snow, especially

7. I. T. Rosenqvist, "Acid Precipitation and Other Possible Sources of Acidification of Rivers and Lakes," *Science of the Total Environment* 10 (September, 1978): 39–49.

near local sources such as smokestacks or feed-lots; (b) gaseous air pollutants such as ozone may damage or kill plants; (c) vegetation, especially on hilltops, may comb polluted water drops from mists. Fog, originating near the ground where the atmosphere has the highest concentration of pollutants, can have much greater acidity than rainfall, which originates higher in the atmosphere. Hoarfrost formed during a dense fog near an industrial town in England was found to have 50 times more acidity than the average amount for rain collected at the same location.

3. Insects or plant diseases.

4. Direct damage from acid rain: while there are no research results indicating that acid rain has a direct harmful effect on forest vegetation at the level of acidity in precipitation occurring in the United States, juvenile and reproductive stages may be more sensitive to acid than mature plants. Studies are now under way to determine the effects of acid rain on seed production, germination, and seedling behavior.

The assessment of damage to the forests and the search for causes of the deterioration, if it turns out to be a long-term trend, is likely to take a long time. The situation will probably turn out to be far more complicated than the oversimplified viewpoint that the forests are mainly threatened by the effects of acid rain on soil.

At the request of the Soil Science Society of America, the Council for Agricultural Science and Technology evaluated the effects of acid rain on agriculture, forestry, and aquatic biology. As of June, 1984, the council's conclusions about effects on soil were as follows:

It is not likely that the additional acidity supplied from strongly acid precipitation would cause a significant decrease in pH value [increase in acidity] in many soils. Numerous studies of artificial acidification of soils indicate that the small amounts of acid added in acid precipitation are unlikely to have significant effects on the various microbial activities that take place in soils. . . .

The amounts of acidity developed in cropped soils as a result of natural processes and management practices considerably exceed the amounts received from acid precipitation. For this reason, as well as the common practice of applying crushed limestone to neutralize soil acidity, it is concluded that acid deposition will not have a measurable effect on the acidity of cropland.[8]

8. Council for Agricultural Science and Technology, *Acid Precipitation in Relation to Agriculture, Forestry and Aquatic Biology*, Report no. 100 (Ames, Iowa: CAST, 1984), p. 1.

Moon Walker

11
Strange Lands

What process created a soil that looks, under a microscope, as though it had been produced in large part by a diligent army of minute glassblowers? Lunar soils, which are two to eight inches deep in explored areas, contain a very large number of glass particles in a variety of shapes—teardrops, dumbbells, threads, and so on.

The origin of soil particles on the weatherless moon is certainly not the weathering processes that create soils on the earth. Rather, the glass particles on the moon were made by meteorite impacts that generated terrific heat, melting the lunar rocks. The splashed molten rock material then immediately cooled into the many-shaped glass beads.

The lunar soil particles are incredibly ancient—many are more than 4.5 billion years old. But the oldest sizable moon rocks brought back to earth are some 200 million years *younger* than these extremely aged soil fragments. The fine particles are older than the larger moon rocks because the most ancient lunar rocks were shattered into the fine dust that now covers most of the moon by a fierce bombardment of cosmic debris during the first few hundred million years of the moon's history.

Dirt

If lunar soil were deposited in your yard or window box, could it be used to grow healthy plants? The moon soils, because of the absence of organic matter, have essentially no nitrogen and very little available phosphorus. However, tests made with moon soil brought back to earth have shown that when nitrogen and phosphorus are added, it will support good plant growth.

There are some very strange soils on the earth as well as on the moon.[1] Tundra and tropical soils, for example, are unfamiliar to those of us in the temperate zone, and they require unusual management—as do peat soils, sands, and polders.

Polders

Totally unsuited for crop planting in their normal setting but fertile under more favorable conditions are the soils of the ocean bottoms. The best known of these lands recovered from the enveloping sea by building dikes (levees) and pumping out the seawater are those in the Netherlands. Just hearing the name "Zuider Zee," the formerly open shallow inlet of the North Sea in the central Netherlands, brings back the storybook feeling of the Hollanders, windmills, tulips, and silver skates. Actually, countries that have relied on reclaiming farmland from the sea include Belgium, England, Germany, Denmark, France, Japan, Venezuela, Guyana, Guinea, and India. The lands recovered from the ocean are called "polders."

The energetic Dutch struggled for many centuries to free a large part of their richest soils from being flooded by the sea and by rivers. In the early years of the never-

1. Unusual soils are described in R. L. Donahue, R. W. Miller, and J. C. Shickluna, *An Introduction to Soils and Plant Growth*, 4th ed. (Englewood Cliffs, N.J.: Prentice-Hall, 1977), pp. 40–41, 463–89, 491–527.

ending battle with the ocean, they faced almost impossible difficulties, as the work of building dikes and pumping out the seawater had to be done by hand. In the fifteenth century the windmill became an important tool in the patient reclamation of the flooded soils. The prevailing westerly winds from the sea kept the windmills revolving, and the windmills in turn drove bucket chains that drained water from the soil and poured it into elevated drainage canals, which sent the unwelcome water back to its home in the ocean. Now, only a few of the picturesque windmills are left as tourist attractions, and the pumping is done by motors. Modern machinery has also made it much easier to build dikes. In 1932 that famous Zuider Zee dike was completed, allowing the reclamation of 550,000 acres.

Building dikes and pumping out seawater was only the beginning of reclaiming the land, however, because salt had to be removed from the soil by leaching with water applications. Gypsum was often used to help remove the sodium salts. Sandy soils could be desalinated in several weeks, but it took several years to leach enough salts from harder-to-drain clay-type soils before they could be used successfully for crops. Deep-rooted plants were usually grown first to help open the soil; rice was also an early crop. After a few years, nitrogen-fixing crops such as beans or peas were planted, followed by grains. Eventually, Dutch farmers could grow all climatically adapted crops on the reclaimed land.

But the battle is never completely won, as excess water seeping under the dikes or from too much rainfall is a constant threat and necessitates extra pumping. And after the water and salt are removed from the reclaimed soils, the land surface drops in elevation for several years as the soils compact and consolidate. If drainage lines or canals are not planned with this prospect in mind, they may easily wind up much too high.

The biggest and most dreaded threat, of course, has always been a break in the dikes. The dikes are built by constructing two parallel walls of clay and filling the core with sand. The side facing the ocean is protected from wave erosion by mat of willows, rice straw, or other material, and covered by rock or cement. One of the worst disasters in Dutch history took place in the stormy winter of 1953 when winds pushed the sea over the dikes in a hundred places, flooding one-sixth of the country, destroying almost 500 farms, drowning 2,000 people and thousands of farm animals. But more than once, the floods have been of vital help to the Dutch when they have opened their dikes to defeat invading armies. In 1574, William the Silent ordered the dikes cut; the action saved Leiden, the manufacturing center of Holland, from a Spanish siege by flooding the surrounding land and allowing a fleet to sail to the city's rescue.

Peat Bogs

Some of the most productive agricultural soils in the Netherlands, as well as in many other countries, are those in reclaimed peat bogs that have been drained, cleared, and leveled. If felled trees and brush are burned in the clearing process, it must be done when the land is wet; dry peat can catch fire, and its smoldering is very difficult to extinguish. Lime is added to reduce the excessive acidity of bogs, and because most organic soils are (or become) deficient in that important trio of plant-available nitrogen, potassium, and phosphorus (and also some trace elements), heavy treatments of commercial fertilizers are generally needed for good yields. When soils made up of fine organic material have been cultivated for some time, they usually become loose and powdery and very susceptible to wind erosion, so a roller or other packing device is then needed to compact loosened organic soils.

The best known bog crop is cranberries, but almost any climatically adapted crop can be grown in good reclaimed peat soils. Once the bog is freed of stumps, tree and shrub roots, and hummocks, it is ready for all sorts of vegetables—both leafy and root types—with celery, asparagus, potatoes, and onions being perhaps the most important, plus such specialized crops as peppermint.

Once peat soils are drained, they begin to settle. This happens because when the water level is lowered, there is a large increase in the activity of decay microbes, many of which require an abundant supply of the oxygen then readily available in the aerated upper layer. The increased rate of decomposition causes much of the material in peat to be converted into carbon dioxide gas, water, and other gaseous and dissolved decay products. As the amount of material in the peat is reduced by the loss of these decomposition products, the fields fall below their original level. In 1848 a long iron post was driven into cultivated peat soil near Peterborough in Great Britain so that its upper end was flush with the surface. After little more than 100 years, the pole projected to a height of 14 feet—even though the land around the post was taken out of cultivation in the early twentieth century and allowed to revert to its former bog condition. Nearby areas that have remained under cultivation have probably suffered even greater subsidence.

What processes in nature create the peat that is so useful for agriculture, nurseries, greenhouses, lawns, and even as a fuel supply? Typically, nutrient runoff from surrounding uplands encourages the growth of many aquatic plants around the edges of a pond or lake. Dead plant matter sinks into the water and accumulates at the bottom of the pond. The water acts as a partial preservative, cutting off the air supply to many microbes, but some slow decay and humus formation are carried out by types of decay microbes that can live and function in the aquatic environment. Over a long stretch of time, the organic debris—along with some transported mineral soil—fills the pond. Eventually, trees and shrubs may cover the site.

The bottom layers of a peat bed are usually made up of the decomposed remains of relatively deep-growing aquatic plants—water lilies or pondweed. This material makes a very bad agricultural soil because it is rubbery; it also dries out very slowly, but once dry, it is very difficult to wet and remains in a hard and lumpy state. In most cases it occurs well down in the bog and does not appear on the surface that is cleared for cultivation.

As the pond begins to fill, the shallow-water plants such as sedges, reeds, or cattails become the main contributors of organic matter. Mosses usually appear where the surface of the accumulated peat material becomes so high that it rises above the water level. These mosses and shallow-water plants are the source of the fibrous peats that generally lie above the undesirable deep-water-plant peat.

Because trees and shrubs finally may become established on a bog, the top layers are often made up of woody peat. This peat is brown or black when wet, loose and open when moist or dry. Woody peat has a lower water-holding capacity than the fibrous peats, so it is not as good for moisture control in greenhouses and nurseries, but woody peat does make a very superior field soil for growing vegetables and other crops. Unfortunately, in the United States woody peats are limited mostly to Wisconsin, Michigan, and New York.

When the materials in a bog are not fully decayed, so that the original types of plants can be recognized, they are called "peat"; if they are so decayed that the original plant parts can't be identified, they are termed "muck."

If there had been no ancient peat deposits the Industrial Revolution might never have begun. Coal, the fuel that made possible the development of industry, was formed from peats that were buried by sediments, compressed, and thoroughly decomposed. The more deeply the peat was buried in the earth, the greater the temperatures and pressures, and the harder the coal that was formed. The hardest coal heats the most, smokes the least, and holds it shape the best, making it easier to store.

Tundra Soils

The decay of organic remains is slowed down not only by waterlogging but also by cold (as your refrigerator preserves food), resulting in many peat and muck accumulations in cold climates. Surface peaty deposits, so common in the tundra soils located in and near arctic and antarctic regions, reduce the transfer of heat from the sun downward into the soil. Thus permanently frozen (permafrost) layers exist at shallow depths.

When the vegetation and peat cover are removed by construction activity, vehicle traffic, or burning, this insulation is lost; the permafrost begins to melt; and the meltwater moves downward to melt still more permafrost. The landscape then becomes a devastated scene of sunken terrain. Along the route of the Alaska pipeline and the hauling road, sunken ground was a constant hazard; it had to be countered by stabilizing the soil with adapted grasses to provide the missing insulation. Areas of off-road driving on the tundra also create unsightly sunken trenches.

Have you ever heard of a "pingo"? Permafrost causes some unusual and fairly spectacular happenings in

PINGO- a giant frost heave with a core of ice, coated
with lichen-covered earth

tundra soils. It blocks downward percolation of water from surface soils thawed during the brief arctic or antarctic summer. The thawed soil may become so wet that it flows gently downslope. With the onset of winter, the thawed layer then begins to freeze from the surface downward, commonly trapping a layer of saturated soil between the frozen surface and the permafrost layer below. Because water expands in volume 8 to 9 percent when it changes from liquid to ice, the freezing surface material exerts great pressure on the trapped, still-liquid soil layer. The soggy material may then erupt, bursting upward to form soil blisters—some as much as 50 feet high; these are known as pingos, which somehow seems an expressive name for them. Water may be ejected from the blister, forming a spring that freezes into a large surface ice deposit. If a pingo erupts in a settled area, buildings may be tilted or engulfed with ice. Trees slanted by the upthrust mounds are called "drunken trees."

The movement of tundra soils by cycles of thawing and freezing often causes a rough sorting, with finer material being moved to lower parts of the landscape and thin soils with coarse rock fragments remaining on the hills. The frost heaving also moves many stones of various sizes to the surface, where they may be distributed sporadically. But over large areas of the tundra, patterns form by continuous alignment of stones in circular, polygonal, or linear shapes. This is a curious sight, and the processes that make it happen are not well understood.

Formation of new soil in the tundra is slow, being essentially limited to the summers, when the upper layers of soil thaw. Because in general the rate of chemical reactions is doubled for every 18-degree-Fahrenheit increase in temperature above freezing, the rate of chemical weathering of rock minerals can be at least eight times faster in the tropics than in the arctic and antarctic regions.

Tropical Soils

The early European explorers who saw the lush tropical rain forests concluded that they must be underlain by very fertile soils. However, natives had learned from centuries of experience that the inherent fertility of numerous tropical soils could soon be exhausted by growing crops. The old, long-undisturbed areas of the tropics—and they are many because the tropics escaped the glaciations that caused so much disturbance in temperate regions during the ice ages—have been so intensely weathered and leached that most of the original minerals are decomposed, and the soils are generally low in plant nutrients.

The luxuriant growth of indigenous vegetation on highly weathered tropical soils is made possible by a unique and awesomely efficient system of recycling. Plant nutrients are retrieved from lower soil depths by deep-rooted tropical plants, dropped onto the surface in plant litter, and then quickly released into the soil in a plant-available form by the rapid decay of organic matter. Removal of the natural vegetation breaks the cycle, and the strongly weathered tropical soils soon lose their fertility.

The native crop growers have commonly used a system of "shifting cultivation," in which they burn off the tropical vegetation, grow crops for two or three years until the residual soil fertility from natural vegetation becomes exhausted, and then shift to another location, allowing the vegetation to return and accumulate more plant nutrients at the first site. Eight or ten years or more after the natural cycle has become reestablished, the natives may use the area for crops again.

Attempts to farm very weathered tropical soils with modern farming methods have sometimes failed and sometimes succeeded. After World War II the British tried mechanized farming for their "Ground Nut Scheme" in Tanganyika, resulting in catastrophic erosion of cleared soils and a $100 million loss for the project. But when properly managed with the addition of nitrogen, large amounts of phosphorus, some potassium, and other nutrients, tropical areas have produced heavy yields. In Brazil and central Africa the plantations grow bountiful crops of bananas, sugarcane, pineapple, rubber, coffee, and cacao.[2]

A common misconception about tropical soils is that *all* of them are highly weathered and relatively infertile for agricultural use. Besides the many very weathered soils, there are others fairly rich in plant nutrients: for example, soils formed from basalt rock on slopes, where erosion will not allow the soil to remain long enough to become extremely old and weathered. Young tropical soils formed from recent river deposits or from volcanic ash also generally have adequate plant nutrients.

Another often-expressed belief about soils of the tropics is that they generally have layers near the surface that irreversibly harden to a bricklike consistency after exposure by erosion or some other disturbance, making the area thereafter unsuitable for plant growth. The formation of bricklike layers does occur, but only in a small proportion of tropical soils, and only in those regions that have strongly developed wet and dry seasons. The powerful weathering in the tropics releases large amounts of iron from rocks. The freed iron combines with oxygen to form oxides, which produce the

2. Gordon Wrigley, *Tropical Agriculture*, 4th ed. (London: Longman, 1982).

red colors so typical of tropical soils. Where the iron becomes very concentrated and then is exposed—as by the clearing of forests and erosion of the overlying soil —the iron-rich material, called "plinthite," hardens into ironstone (also called "laterite") by repeated wetting and drying.

People in Kerla State, in southern India, discovered more than 2,000 years ago that they could farm soils that contained plinthite as long as they never exposed the plinthite layer to sun and rain. They grow many types of herbaceous plants, including vegetables, as well as woody types, such as fruit trees. By planting them together, with the lower herbaceous plants beneath the woody plants, they provide consistent protection against erosion. Plowing is avoided, and planting and harvesting are done entirely by hand.

Plinthite, while a severe threat to agriculture in some areas, can be a very useful building material. People who live in regions where plinthite is plentiful can expose it and cut it into bricks while it is still doughy in consistency. The bricks harden into ironstone after a few months of direct exposure to repeated wetting and drying by the rain and sun. Some clever tropical residents have even cut well casings or statues from plinthite and then allowed them to harden.

Beautiful specimens of ironstone exist in some currently very nontropical areas—near Lincoln, Nebraska, for example. Others were discovered under the soils of the Sahara, where ironstone could not form now. These finds of ironstone in unexpected locations show how different the climates of those places must have been in the past.

There is generally less sand in the tropics than in the temperate and arctic regions because weathering is so strong in these hot, humid climates that it causes sand grains to decompose into finer material. Sand-sized particles in the tropics are often composed of smaller particles cemented together by iron.

Farming on Sand

In some areas where cropland is limited, such as India or Great Britain, expensive attempts have been made to improve sand deposits by mixing in clay soils. The finer particles help improve the sand's water- and nutrient-holding abilities. Organic matter can also be added to improve fertility and water retention, but it decomposes so rapidly there that it is difficult to maintain more than 1 or 2 percent of organic matter without heavy seasonal additions.

Is it possible to grow crops on a sand dune without extremely expensive treatments to change the nature of the soil? It has been done: near Pasco, in southern Washington state, dunes were graded to slopes of 5 percent, and water was piped in from the nearby Snake River for irrigation. Wind erosion was controlled by keeping the land almost constantly covered—growing crops like potatoes or wheat, and leaving the stubble for ground cover. Frequent water applications with sprinkler irrigation kept enough water available, and repeated additions of fertilizer in the irrigation water maintained the soils in a fertile condition.

The establishment of moisture barriers is another method that can be used to retain sufficient moisture in sands. In Michigan, Israel, and Taiwan, an ingeniously designed machine has applied moisture barriers by (1) lifting an area of sand eight feet wide with a sweep; (2) applying liquefied cold asphalt to the exposed lower sand; and (3) restoring the displaced surface sand. The impervious asphalt barriers have increased yields by 50 to 100 percent; they pay for themselves in three years and last for seven years.

Index

Dirt was composed into type on a Compugraphic phototypesetter in ten point Melior with three points of spacing between the lines. Melior Bold was selected for display. The book was designed by Cameron Poulter, composed by Metricomp, Inc., printed offset by Thomson-Shore, Inc., and bound by John H. Dekker & Sons. The paper on which this book is printed bears acid-free characteristics, for an effective life of at least three hundred years.

TEXAS A&M UNIVERSITY PRESS : COLLEGE STATION